KOSOL
AND
KOEUN NOUN OUCH
SPHERICAL
GENERATOR

KOSOL AND KOEUN NOUN OUCH SPHERICAL GENERATOR

(A Path to Free Energy and Hyper Drive Technology)

by
Kosol Ouch
Koeun Noun Ouch
and
David Lowrance

E-BookTime LLC
Montgomery Alabama

Kosol and Koeun Noun Ouch Spherical Generator
(A Path to Free Energy and Hyper Drive Technology)

Copyright © 2006 by Kosol Ouch, Koeun Noun Ouch and David Lowrance

This document is part of the **[public domain]** and is intended for mankind to improve our overall "collective consciousness" leading to the use of clean energy. The writer assumes no liabilities, has absolutely no concern for monetary gain, places no limits on the use, copying, or distribution of contained information, and deserves no credit for anything other than mastering comprehension along the way. Playing with Scalar waves or high voltages can be dangerous. The reader is to assume all liabilities for applications of principles found within this document. For the greatest good of all involved!

Vince did the circuit drawing and Lacosta did the 3d model.

It is now time to start building more devices!

Library of Congress Control Number: 2006925404

ISBN: 1-59824-192-3

Published April 2006
E-BookTime, LLC
6598 Pumpkin Road
Montgomery, AL 36108
www.e-booktime.com

I would like to dedicate this to everyone involved, especially Kouen Noun Ouch, Lacosta, Vince, David Lowrance, Jake Tepac, Jesses Sanchez, and Carlos Sanchez.

A Theory of the Universe

The Forces of Nature

Based on Wilbert Smiths New Science 1950's, [spin and relativity], the Russians study of Torsion fields, the observations of the recorded works of Searl, the Hamel cones, Floyd Sweets VTA, Tesla, Keeley, Tom Beardens description of Scalar waves, and the Kosol Spheres now being developed.

This work represents a compression and combining of much previously presented **discussion** and **theory** on the subject of alternate energy, gravity, and time brought into one place. The intention is to perfect the Kosol device, now seen by the authors to be the most complete model yet offered for the mastery and use of alternate energy.

Contents

FOREWORD 9
A REVIEW OF THE CONVENTIONAL MODEL 13
   The 4 Forces in modern physics 13
   Comment 14
SECTION ONE [A New Physics] 15
   Background Info 15
   Field Theory 15
   Relativity 17
   Fields 19
   Chemistry a new model 26
   Neutron Theory 29
   NMR [Nuclear Magnetic Resonance] 32
   ESR [Electron Spin Resonance] 32
   Electric Arcs 33
   Tilt or Wobble [Precession] 34
   Induction of Gravity 35
   Spirals 36
   Gravity Control 37
   Forces Compared 39
   Theory of Gravity and Aether 40
   Cold Energy 46
   Planar Angle of Spin 50
   Occult Chemistry 52
Section Two [Devices] 54
   Scalar Canceling or Tensor Devices 54
   Tom Beardens Scalar Potential and the Vacuum 56
   Legend of the Circle Antenna 57
   The Delta T Antenna 59
   The Three Fields of AG 61
   Spinning Magnets 63
   The Tornado Effect 64
   The Roschin and Godin Searl Duplication 66
   The Marcus Device 67
   Magnetic Gating Technology 68

Contents

    Spinning Metal Cylinders, Spheres, and Pyramids........ 69
    Possible Helpful Awareness About Working With Such
    Devices.. 69
Section 3 [The Kosol Device Platform]............................ 71
    What is a Kosol Spherical Device: 71
    Field Generation with a Sphere...................................... 72
    Magnet Patterns 75
    Scalar Magnet Patterns on the Kosol Spheres 77
    The Core Magnet Electron Accelerator 80
    Kosol Sphere Analysis... 81
    Field Considerations 82
    Spherical Torus Field [Magic].. 86
    Intensity and Effects 92
    Construction... 94
    Notes and Suggestions to Expand Awareness 98
Section 4... 101
    A Search for the Perfect ZPE Device 101
    More on Magnets 103
    Post Analysis... 103
Electrical and Mechanical Analysis for Kosol Device ... 105
Closing Comments.. 113
ILLUSTRATIONS 116

Foreword

A study of Wilbert Smiths work leaves one with a new view of relativity where all time flow becomes a function of distance from center of spin in a universe filled with separate densities, each one a new subuniverse. Smith introduces us to a universe where UFOs travel across [density] in order to traverse large distances then pop back into this one in the blink of an eye. A study of the Russians work with Torsion fields leads one to interchange Smiths "Spin" with a new label, simply "Torsion" bringing the observed interactions to a very recognizable and basic force of nature. A concept that has already shown weight reduction over 100 years ago in large heavy wheels that were spun up, hit with a hammer to "precess" and then easily lifted with one hand.

It has long been theorized that gravity is emitted from matter. It is my observation that gravity and time are so interlinked that they very probably originate at the same source, inside the atom. With time being viewed as a torsion force it becomes possible to develop a general intuitive theory to begin to explain some of the reported effects displayed by the SEARL disk, and many other devices claimed to alter gravity. Gravity and time may share a quadrature relationship just as electricity and magnetism but operating at a much higher frequency. A stronger gravity may simply be the result of the atoms smallest particles spinning at a lower rate and generating a slower time field for a local area of space. In this sense it is not space that is curved by gravity, but gravity that is curved by time. While it is hard to envision a single atom emitting a force capable of reaching out through the universe as far as gravity and time, it is not so hard when you realize that gravity does do this. If time is linked and follows the same path as gravity then the force powering the universe from

Foreword

second to second is truly found within the atom, a unit that never needs recharging and never runs down. Energy may be shifted between the gravity and time fields just as it has been done with electricity and magnetism.

My connection with Mr. Kosol Ouch has been an inspiration on this project from the very start. His method of "connecting the dots" was invaluable in combining the different disciplines necessary to see a larger picture. The truth is stranger then fiction to which this work is a living testament. After my experience of personally connecting with the fourth density aliens using Kosol's guidance, I was left with two messages lingering in my awareness for many weeks while initially working on this project. **Comprehension** is necessary for man to reach the next level. We are responsible for our advancement, and only with personal **comprehension** at each step will we be able to create the next generation of devices powered directly off the universes most primal force. The second message was the importance of discovering **"the five"**. The five still remains a mystery to me at present. Whether this meant 5 principles, 5 elements, 5 energies, or 5 people I have yet to discover.

My own addition to this work, that of recognizing, as quantum physics has shown us, that as frequency raises, so does energy. As we move towards the center of "spin" through Smiths higher densities, while vibrational rate increases, we approach not a **dead center** but one with **unlimited energy**. The ultimate or "God force" is seen brilliant and powerful, [**Source**] where vibrations are far accelerated. This is a merging of an ancient spiritual concept known as the TAO, but is also found within a current medical instrument, the MRI. All matter is now seen as a dance of opposite forces around ultimate awareness, [God]. Electrons are seen crossing the threshold of our density as they wink on and off, carefully opening and balancing the forces of this universe through all densities of higher vibration. The inflow and the outflow of Source manifesting as torsion [time], gravity, electric, and spinning magnetic forces. This work is dedicated to the understanding and comprehension of these forces, and the

Foreword

devices said to alter them. It begins with theory and then moves on to the actual devices available for study at the present, attempting to offer a workable model for ongoing development of the Kosol Spheres.

A Review of the Conventional Model

The 4 Forces in modern physics

The strong force:

Said to be the force holding the nucleus of atoms together, holding protons and neutrons together, reaches out magically to the edge of the atoms nucleus and then disappears at 10^{-15} meters. Gluons are said to be the mechanism of this glue like force, very powerful and very short range. Neutrons do not have a charge and therefore cannot be manipulated like protons and electrons. Neutrons die after being removed from the nucleus for 15 minutes. They break down to stable protons and electrons giving off another imaginary particle to make the numbers add up. This force has not been connected to gravity or time.

Strength = 1, range = 10^{-15} meters

Electro Magnetic force:

1/137 as strong as the strong force, electro-magnetism has no limits on its range.

The Weak Force:

The neutrino interaction which induces beta decay.

Strength = 10^{-6}, and range = 10^{-18} which is 1/1000 the diameter of a proton

Gravity:

Said to be the weakest force at a strength of 6×10^{-39}, with an infinite range.

Comment

This is what modern science has come up with after having studied the decay of matter over the past 200 years. As well as a host of other particles to explain each process observed in atomic decay, attempting to make it fit the formulas of previous "immutable laws" known within other formulas. With the focus on study of "decay" of matter, and a postulate that all the energy in the universe was released at the big bang. The universe is seen as an explosion slowly winding down and dying. No regenerative force know as [Source] has been located or included within modern physics, and no cyclic renewing of our universe has been identified that would support a theory of ZPE. If the universe is in fact slowly decaying away, [entropy] will ultimately win and it will either end as a lifeless expanse, or re-collapse back in on itself.

Although gravity is said to be the weakest force, it is also said to bend or warp space and time creating a seeming paradoxical universe where the weakest force of nature somehow has the power to alter spatial dimensions. I would suggest that we may simply be measuring gravity as one of its "effects," that of matters attraction, rather then correctly identifying its true nature linked to time as a quadrature relationship. The weak force we classify as gravity may be only a very small difference between two very powerful opposite forces of the fabric of space and time.

Section One [A New Physics]

Background Info

To date I have found several approaches to Gravity control, but mainly two flavors, that of Aether Theory and that of Field Theory. No collection on the subject would be complete without a presentation of both. I shall begin with the consideration of Field Theory and then build towards Aether theory mixing in spiritual knowledge and creating a vocabulary from all disciplines. The goal is **Comprehension**.

Field Theory

Source:

Consider that to have awareness, there must be a [**Source**] a beginning, a point where all awareness, energy, and matter are brought into this universe.

Space:

All fields function inside space as we are aware of it, and thus an understanding of dimensional space is necessary as a first step. Space is made up of 3 basic yet separate elements of awareness which all build on one another around [Source].

Line:

A line in space is the stretching out of a single point from source in one direction. Within the consciousness or awareness of a line exists nothing spatial or unique, merely a direction impossible to distinguish from either side, or impossible to recognize or even agree universally on any fixed measurement. All measurements on a line are arbitrary to any universal cultures perceiving them.

Area:

A line from source stretched out to form an area, now has the qualities of a surface which exist in a plane. Directions and increments are still arbitrary.

Volume:

An area stretched out to form a volume, now possessing something real that our awareness can perceive and "intuitively" comprehend as space. Each of the above elements relies on the previous ones for its existence. None can exist without all the preceding ones.

Spin:

At this point of having only 3 D space there is still no measurement possible to which any culture having awareness can agree. All measurements are arbitrary along any line, area, or volume. It is only with the introduction of spin that a concept of exact universal measurement can be observed by awareness. Consider at the center of your local space a line, at [**Source**] is stationary on one end, rotating through the area of a circle. We now have the first universal measurements developed. Angel, curve, rate of change, period, frequency. Measurements can now be present which all awareness must agree on, thus they are concrete realities for all awareness existing in 3 D space. 1 rotation is a finite quantity. 1/4 rotation is also finite. 90 degrees etc. Our labels can vary but the "intuitive" principle is exact.

Now consider the center of the circle represents [**Source**] the center where all energy enters into our universe, having qualities of infinite energy where awareness is perceived operating at light speed.

Relativity

Simultaneous spin, Density, Light Speed:

Consider two points. Point 1 is 1 unit from [**Source**], point 2 is 2 units from [**Source**]. Energy within Electrons and Protons at point 1 has a higher energy then point 2 because they are closer to Source. Source energy is infinite. They both begin their journey around center at light speed and arrive back at starting point simultaneously and agree that 1 rotation has expired. This is the only way both could agree. Point 2 [seems] to have traversed twice the distance. We can see that point 1 time scale must be faster relative to point 2 and its distance covered more [**dense**] to hold more energy in less space. Point 1 [**light speed**] must be lower then point 2. If a linear function was applied the light speed at point 1 relative to point 2 would be 1/2 c. It would seem that at [**Source**] light speed would halt, and you would simply become the "light." Now [light speed] becomes a function of how fast our awareness is accelerated towards it, and [we] are the ones seen moving not it. If light speed has lowered towards center for us then linear dimensions have been compressed and space in this region contains more energy. As respect to point 2, point 1 is more [**dense**]. Point 1 relative gravity and momentum are lower then point 2. While closer in to [**Source**] we can move much faster around the universe, and people in outer densities will appear to freeze because of rate of light speed change relative to us. Some point along this line will represent any region of space where change appears to happen at a constant rate with respect to light speed. Thus resolving the paradox of Einstein's dual relativity, by referencing all points to center or [**Source**] rather than referencing them randomly to each other. As awareness moves inwards between densities, towards Source, it is moving closer to light speed, so light speed appears to slow down, yet all matter holds more energy.

So spin is seen to have geometrical qualities as an angle, period, or frequency and should be thought of as a constant. Time [Torsion] is the variable not spin, this places a new universal reference between densities and the speed of our awareness as it is distanced from source becomes a variable. So light speed is no longer fixed with respect to us.

Density Threshold:

The lower or upper distance from Source where matter appears to fade out crossing into another density and disappearing from view. It has been estimated that as much a 98% of the matter in the universe is invisible to us as [Dark Matter].

DS factor [distance from source]:

Within relativity, the factor is [Distance from Source], being 0 at source where Light speed is 0 relative to awareness and consciousness is God Awareness.

AW factor [Awareness factor]:

Relative speed of awareness to light speed. At DS = 0, AW is infinite.

Time as an increment within spin:

This is the increment of spin for any one distance from Source. It has a linear quality existing in three space in any local density. Its analogy to a line is "intuitive" from our perspective, moving normally in one direction. Light speed, although changing between [densities] with respect to awareness, is a constant range for any one [density]. Time allows us to increment universal spin based on our local [gravity-torsion] and our relative [light speed.]

Density:

Consider density forming an ordered structure, grouping itself into bands or rings of distance from source where realms may exist separately which become invisible to one another. Earth is said to be presently in the 3rd level of ordered density moving towards the 4rth. Looking at a 4rth density realm which is moving closer to light speed with a lower DS, they may appear to be able to move very quickly and fade in and out of our density by increasing their DS. For us to see them we would have to decrease our DS [distance from Source] raising our AW [awareness factor].

Source is everywhere:

A spiritual concept, assuming an intelligent quality. Consider that [Source] may move anywhere in the universe and begin to create reality on any density. A connection to source is necessary to maintain what is created and keep it relative to all other creations. Matter or energy moves in both directions through this connection creating the dimensions that exists around it as well as the local space time fields for the object that may now interact with other Source connected objects to build a local universe.

Fields

With this background in space we are now ready to consider the known Fields.

The primary fields:

Torsion:

The emergence and withdraw of expanding and contracting energy from Source, having a vector of "perpendicular to spin direction" with directions reversing. The flow of Aether, having elastic qualities at high RPM and solid qualities near light speeds. Torsion is radiated off a spinning wheels ends when the wheel is given a movement perpendicular to rotation and becomes a [Torsion field.] The main torsion of the universe is radiated off the smallest movements of atoms and creates the time field. "Spin" being the only universal concept able to establish a consistent relativity across the universe to all awareness perceiving it.

Torsion [Time] Field:

Torsion is a 1D filed moving into space with a direction vector generating the time field for a given density. All being relative to [spin] and time being a variable dependent on awareness's distance from source. As the atomic particles spin and also orbit they will spread this field out spherically as a 1 D field.

The first field necessary to hold all others is one allowing change. All frames of reference are altered by changing torsion including relative light speed. Torsion has linear qualities. As spin fills space at distance from source, the Torsion field is created across all densities and is relative to distance from source. It has been speculated that all matter in one local region of space, in order to attract and form higher structure, must have the same Torsion field. Planks constant may represent the spin present in our density, which is claimed to be constant for all matter. Violent fluctuations in the torsion field will cause matter to loose its cohesion and break apart. All tensile strength relies on a coherent torsion field presence.

Gravity as a 2D field:

The second field necessary to begin building a realm within a density. A force that can reach out and interact at the greatest distance. Manifesting 2D qualities and falls off as a function of inverse distance squared. Gravity must work for atoms and galaxies alike, therefore must emerge from the smallest creation of nature. Its source is in the atom. Gravity is a strong force which may be a balance between two opposite radiated 2D fields. One expanding outwards to fill space [antigravity] and one compressing inwards towards source [gravity]. Gravity can be thought of as the area of a circle as it expands and then contracts along a torsion field into and out of source. Gravity shares a 90 degree relationship to torsion. [It is suggested that the expanding force radiates from the AntiGravaton, the contracting force from the Gravaton. The two particles emerge from the tornado of source doing a dance of balance. When torsion is moving in a straight line inflow and outflow are equal and cancel. When torsion is curved into a circle of atomic proportion gravity becomes predominantly inflow or outflow. Combining the Proton, Electron, and Neutron interactions gravities overall force is one of pull towards Source. This is due to relativity as viewed from the 3 point reference. As the gravaton pairs rotate around a helix they share different times within our density. Based on direction of spin and dominant vortex particle presence.

[Gravity - Torsion] Field:

As the Gravity 2D field moves through a vibrating circle of the atom it inscribes the pattern of a spherical torus around the atom. Gravity is seen to operate from its center with a spherical nature falling off equally in all directions from center. Gravity has the shortest DS [distance from source] thus is the strongest force working inside a torsion field. May be thought of as radiating from a pair of gravatons in a tight helix having curved torsion and area moving through space. Torsion is the line through the center of the helix giving it direction. As a 3D field it falls off as a function of distance cubed and has a magnetic quality in this mode. Bundles of spiraling pairs of gravatons and antigravatons rotating in a circle around an atom form electrons. Electrons have a negative electric effect and gravatons may be balanced and act differently when placed into a curving path of altered radius. Gravity operating as a 3D field within the atom may be the force that keeps Electrons from crashing into the nucleus of the atom through reversed induction.

None of the above fields can exist without all the previous ones to build from.

The secondary fields:

Electric potential:

Static:

A static charge, extending its force outwards in a spherical pattern in space having a fixed gradient in space but moving through time. This field is not in movement but stationary. It can be pictured as a bundle of Graviton pairs spinning with no curved torsional movement through space, therefore gravity is balanced to 0. Its field is "intuitively" compared to "area" in space, as its field strength falls off as a function of the distance squared. Electric field falls off as an area or gradient decreasing outwards from the center of a sphere in space. Consecutive hollow spheres of larger radius will have a lower electric potential. The capacitor, an electric storage device, is designed around two geometric planes, very close together, an "area" type device where a gradient measured as voltage may be stored. Static electric field has the quality of stored pressure where no relative motion is present in 3D space. Electric potential appears in positive and negative spheres of field which attract one another. It is a weak force compared to those coming before it and can not exist without them. May be thought of as a bundle of spiraling gravatons and anitgravatons. The increase or decrease of potential may be thought of as a [different distance from source] for the gravaton pairs spinning inside it. In the case of the proton the direction of spin is reversed. Thus charging the air with a Van De Graph or Telsa coil adding a negative potential tends to decrease the distance from source for the overall system. Part of Hutchisens work observed.

Current, a moving electric field:

As the electron is moved through space with the introduction of a positive field to attract it, it moves straight, therefore the gravity field generated is in perfect balance and results in 0. The electron is now moving through a wire at a constant [voltage] indicating its inverse distance from source. Current is present and may be calculated using all the present electronics theory. If no curve is present in its path at near atomic dimensions then no dominant gravity field will manifest. The electron is seen to be an over all [expanding force] and the proton a [sucking force]. Thus electrons may be producing the overall Antigravity field, and Protons the overall Gravity field. Doing this in tight atomic curved movement, due to their opposite spin].

Note:

Interesting to note that almost all our theory on electronics is from observing the movement of electrons, and little is known about the nature of positive electrical fields. It is observed that static negative electric fields repel, and that electrons are attracted to positively charged atoms. However it may not be so clear what happens between positively charged atoms. As electric current through a wire is always the electrons moving it may be hard to know for sure.

Magnetism:

Basic:

When an electric field, having the quality of area [an electron], is moved through space, and time, it produces a volume. The volume produced by electric potential moving through space and time is magnetism.

A magnetic field can be seen as a circle spiraling around the electron as it moves straight, it is the [curve] of moving electrons, felt as a force over distance.

Magnetism falls off as a function of the distance cubed, as a volume expanding away from the center of a sphere. Magnetism is present within a volume, as with a permanent magnet. In an electromagnet the force of a magnet is felt to be pushing 90 degrees offset to the electric fields direction of motion that produces it. When the direction of electric field is moved into a **circle** or **spiral** coil, magnetism takes the shape of a spherical torus, one end of the coil takes on a dominant north pole and the other a south pole. The magnetic field strength is a function of electron **current**. This is the second field exhibiting this quality and is naturally present in the atoms of iron, cobalt, and nickel due to there atomic structure.

It is suggested that magnetism is the result of the gravaton pairs helix curve that is on a larger diameter circle, IE a spiral like gravity with a larger magnitude of radius. Thus the forces source is the same as gravity however its magnitude and frequency are lower due to increased radius [distance from source]. The radius of magnetism is seen as the distance across complete atoms, or in wires of much larger radius curve in electronics, while the radius of gravity is seen as the distance to helix center for a very small movement of the electrons spiraling paired particles.

Complex:

There is another theory of magnetism that emerges from the study of what torsion does as a large number of electrons move through the shells of atoms in one direction. As a torsion field is radiated off spiraling electrons as they orbit the atom, causing more to move through a spherical shell one direction causes an imbalance of canceling torsion which becomes non canceling at the 90 degree points of the electron shells orbits from the points of electron entry and exit. Magnetism is thus viewed as a bidirectional torsion field in this case originating only from the outer valence shell of the atom. Thus a magnetic field can be seen as a time compression in two directions only. Slightly altering the atoms overall torsion field in only two directions and actually altering the flow of time.

Effects of magnetism on atoms:

A strong static magnetic field induced into an atom is seen to raise the frequency of its magnetic vibrations.

From the study of NMR and ESR, the concept of spin enters into conventional physics. As Electrons spin around the nucleus in an orbit their motion generates a magnetic field seen to precess around the magnetic field that is present. This rate of precession is increased to a higher frequency as the external magnetic field is increased.

The atom has natural precession rates that have been measured as frequencies. Each element is different.

Protons, Electrons, and Neutrons also spin around their center, and also precess around a magnetic field present. Protons precess in the Radio frequency range, and electrons in the Microwave frequencies.

Protons 42.5781 Mhz / Tesla
Electrons 28.025 Ghz / Tesla
Neutrons 29.1667 Mhz / Tesla
Deuterons 6.5357 Mhz / Tesla

Placing a proton in a 2 Tesla field will double its frequency to around 85 Mhz.

By transforming the waves emitted by the atoms to a frequency domain, a spectrographic signature of each element is the result. An MRI used in medical imaging systems can identify most atoms and generate images using this technology. It is based on the precession of atomic particles absorbing and emitting these high frequency photons.

Electron particle spin is opposite of proton spin. Protons are spinning the same direction as their orbital spin but electrons are seen to spin the **opposite** direction as their orbital spin direction.

The above fields, like the spatial fields, also share a 90 deg relationship. Also all the ones before each are necessary for its existence. IE without electricity there can be no magnetic field. Without time, neither can exist within awareness.

At this point we have now [intuitively] suggested a definition for all the forces of nature, within two particles dancing within a yin and yang structure [the electron, and proton] a dance that moves around the atom in its shell of vibration.

Chemistry a new model

[Searching for the source of gravity torsion and space time]

It is now necessary to observe chemistry to see if we can explain the differences inside electrons and protons.

Looking from the outside we see Protons having a positive quality of voltage present based on their distance from source and their dominant gravaton moving in and out of this density. Therefore we may identify them with a dominant Gravity as well. If both an electron and a proton have the same two forces operating within them, then the balance of release must be simply different. That is the Proton is the source of both positive potential, and Gravity inflow. Its larger circular movement causing positive electric charge and its smaller gravaton circle causing Gravity. The electron having a larger orbit and yet a negative overall electric charge, may be the source of over all AntiGravity and not Gravity. This is all a function of Helix curving to move gravaton particles in and out of our density for different times. Thus we have now seen both ends of the threshold moving through our density. Speeds and directions of gravaton particles in both composite particles must all interact to produce the overall Field of Gravity. We know the positive voltage and the sucking Gravity are winning at the moment.

The 4 forces viewed as a [simplistic] helix particle theory:

Gravity, Anti Gravity, Electrostatic [Positive and Negative], from a single model. The helix of [gravaton and antigravaton] pairs spins two different directions. In the one direction it manifests as an electron, in the other a Proton. The two setting within a field stationary will attract. This gives 4 possible combination states for composite particles to produce field forces.

The primary particles:

The primary particles operate at the edges of our density, winking in and out as they move between, delivering either expanding force or contracting force in a clockwise or counter clockwise direction.

AntiGravaton = Particle allowing inflow of energy from source while it is at density threshold.

Gravaton = Particle allowing compressing outflow of energy to source while it is at density threshold.

Gravaton Pair = The two particles emerging from source at the highest energy level of our density where the threshold may be crossed, minimum distance from source threshold in our density. They appear to move in and out of our density and are found spinning opposite direction in composite different particles. The pair consists of an AntiGravaton and a Graviton.

List of composite particles:

Proton: [Source of **Gravity** [inflow of Aether] and **Positive Electric** Potential]

Gravaton pair:

Gravaton spiral entry to our density CW [contracting force] (as viewed from here)
AntiGravaton entry to our density CW [expanding force]

The two particles spin around one another as they emerge and regress across our density threshold.

When this particle moves in a tight atomic radius curved path it produces strong forces of gravity due to its helix being sharply curved. Allowing Gravatons to spend more time in our density.

While setting still or moving straight it would be expected to have a positive electrical charge. Its inverse distance from [source] is positive voltage and should be expected to operate all the way across the [DS factor] found in our density.

Electron: [Source of **Antigravity** [outflow of Aether] and **Negative Electric** Potential]

Gravaton pair:
AntiGravaton entry to our density CCW [expanding force] (as viewed from here)
Gravaton spiral entry to our density CCW [contracting force]

When this particle moves in a tight curved orbit at atomic radius it produces strong forces of AntiGravity due to its helix being tightly curved. Allowing AntiGravatons to spend more time within our density. While setting still or moving relatively straight it is the "electric" electron we are familiar with having voltage potential as its distance from center, which will operate all the way across the [DS factor] found in our density. Electrons orbiting an atom are approaching light speeds .99999999995% c. This is very near [source] from our point of view, thus they may have incredible power to tap ZPE if their motions are correctly manipulated. The AntiGravity filed they radiate is very powerful with atomic sized curves yet weaker overall then the protons.

Summary:

We have now offered a simplified model to explain the four forces at work to provide 4 fields within our reality all based on a flow of Aether pressure to and from source. The strong forces of Gravity and AntiGravity and the weak forces of Positive Electric and Negative Electric Potential. Magnetism is created from the electric field in motion and Torsion from the Gravity field in motion. To get to this point required a new theory of relativity based on distance from [source] and a three point reference. Assuming that all matter exists in an already moving state, thus can only be referenced to other matter with respect to its motion relative to source.

Neutron Theory

There is also a belief that Neutrons may be the source of gravity. Electrons and Protons the source of electricity. This becomes increasingly hard to ignore with the simplistic theory offered previously offering no explanation for Neutrons other then to be a neutral buffer between electron and proton. The scalar Smith coil shows an important principle.

As an Electron and Proton are combined, they form a new stable structure of vibrating opposite energies, the Neutron, which may be the actual gravity producing unit within the atom. The Neutron star lends credibility to this theory as well as the Smith coil. Now separating the forces of electricity and gravity into different particles within the atom operating at different distance from center. The frequency of gravity now looses its vibration across a band of Electron shell frequencies and may become a single 2 D superluminal vibrational component unique to Neutrons. The explanation for the SEARL and Helsinki devices may now have to involve the Protons proximity to the Neutron, and tapping cold Protonic energy having a much greater effect on Neutron vibrations, then Electrons do. Or the interaction effecting Neutrons may be one of a [scalar canceling] nature which mimics its own structure.

The Neutron seen as a scalar cancelling force

Visualizing a Neutron we now see a proton charged + and an electron charged - doing a **scalar canceling** dance, canceling out their EM field and radiating the gravity torsion field, or space time itself. As the two energies are much closer together then when separate in the atom, they would be seen at a much higher energy state. An electron and a proton vibrating in an extremely tight atomic radius, much closer to center of spin then an electron orbiting the atom. This sounds a lot like a scalar canceling coil and may come the closest to the model of Smiths tensor beam. The positive and negative charge of EMF pushed together in an opposing EM field cancelling pattern resulting in a tensor or time beam. Thus the universes source of torsion may be located here within the Neutron. One need only pick up two Neo magnets and attempt to push them together in opposition to break out of the teaching that opposing magnetic fields cancel one another.

The Neutron being a combination of Electron and Proton, may be effected by electric and magnetic forces in very close proximity, however it is more likely that it would be effected mostly only by **torsion** or **spin** or **scalar radiations**. Neutrons because of their neutral EM field are not easily manipulated with EM fields. Therefore when attempting to alter the operation of Neutrons the important factor to alter would be **torsion** or **tensor**.

There is a theoretical threshold where as gravity increases within a neutron star a limit is reached where matter can no longer compress and begins to cross the density threshold. The current theory states that it moves into a "warp" or black hole, however I prefer Smiths view of a density shift towards becoming invisible matter. The matter may be seen as moving into another density where gravity and time are altered. This too may leave us with a new feel for what a Neutron actually is, a stable connection to other densities or a mini black hole.

If this model is correct then the creation of an electron shell in an AG device, will tend to neutralize the protons positive dominance within the atoms where it is contained, bringing the balance closer to the Neutrons EM canceled state. A high negative charge therefore would be seen as an important field to establish an effect, but an EM scalar canceling tensor force may be far more important to effect Neutron spin function. In the SEARL model the positive charge towards the center, the containment field functions to harness the negative electric potential but may offer no other effects in itself. The copper atoms at the surface where electron potential becomes very high may be where the effect is happening. Thus it becomes important to release the scalar or tensor energy within this electron shell, and a positive charged field becomes of little value to pushing matter towards the density threshold.

The model of Neutron gravity moves the strongest forces of time and gravity into the nucleus at the center of the atom. This would also shift the main force of inertia inwards to the Neutron. At this atomic level of operation the Electron and Proton operations are now seen as very weak forces in comparison being around $1/137$ as strong as the force found inside the nucleus.

Normal matter is now seen as a tensor beam radiating outwards from a positive charged center through a negative electron shell. The Searl device is seen to reverse this flow, setting up the tensor at the negative outer electron shell having a canceling or reversing effect of the Neutron.

NMR [Nuclear Magnetic Resonance]

Almost all atoms have noncanceling spin at the nucleus [proton orbital magnetic spin] that will align to an induced magnetic field in one of two possible ways. The alignment has precession, and when the atoms allign they will either fall into attraction [considered a low energy state] or opposition [considered a high energy state] precessing around the static magnetic fields direction.

If the atom is held in a magnetic field and then stimulated at 90 degrees to the field with a frequency between 1 Mhz to 700 Mhz, at the exact **Lamour frequency** of resonance equaling its magnetic **precession rate**, the Proton alignment will flip 180 degrees to the high energy opposing state and absorb a photon. When the frequency is removed it may flip back releasing the photon. The photons will be of the same frequency as the precession present in the Proton. For NMR this is a Radio Frequency, usually 1 Mhz to 700 Mhz and raises with the magnetic field strength, .5 to 4.5 Tesla. MRI units require supercooled coils to produce this strong a magnetic field at a distance, however the Helsinki device had aprox 1.5 Tesla within its cylinder and could have easily reached this ability to flip Proton spin if the correct RF was present. The flipping motion is not a direct instant flip, it is instead a spin path like a bar magnet moving through a conical shape on each of the atoms ends. Both ends of the magnetic field spin wider until they are traversing the equator of the atom, then they continue to spin around to precess around the opposite pole of the atom, reversing the magnetic field of the proton shell as they cross center.

ESR [Electron Spin Resonance]

Electrons flip at a much higher frequency well up in the gigahertz range, but with a much lower magnetic intensity.

Kosol and Koeun Noun Ouch Spherical Generator

Magnets as low a 200 gauze can be used, but frequencies go easily from 2Ghz to 12Ghz. Electrons are 1000 times lighter then protons and precess around a centering magnetic field at a much higher frequency then Protons. This process may be in effect in the Helsinki device as the thickness of the cylinder is supposedly an exact multiple of the ESR wavelength for neo magnetic material and would seem to form a resonant section for standing waves along the X axis of the device. This may be one reason that torsion is moved from atoms to device movement.

If the magnetic field is strong enough then electron spin and rotation come into alignment coupling angular momentum and magnetic field. This is only seen as something happening in the sun through spectography.

The cones found in the Hamel devices could be resonating these same frequencies and help cause the vibrations to surface, as a cone shape may allow a range of frequencies to resonate around its tapered length.

Electric Arcs

An electric arc is seen to give off a wide spectrum of energy as the electrons actually jump free of atoms to form a plasma for a brief instant as they jump between electrode surfaces. It has been noted experimentally if not accidentally that arcs give off frequencies from low audio up through microwave, infra red heat and even ultra violet rays. Electrons forming shells around magnetic fields are seen to glow violet or blue as standing balls of light. It would seem one of the reasons **Tesla** moved to impulse energy leaving sine waves behind was that the nature of pulses of arcing electrons do give off a wide range of spectrum that can be channeled into coils and reclaimed from across a wide band jump to a narrow band. A pulsing DC arc can be used to run a giant Tesla coil.

Tilt or Wobble [Precession]

View from outside sources:

Such a field, as gravity, may be produced from a "precession" of rotation, where top and bottom rotation axis of atoms rotate also in small circles, producing small variances in the spin rate on both ends yet maintains a steady center at the core. The axis of the atom now viewed as the intersection of two cones with a stable center. The actual spin rate has not changed however this circular movement viewed from outside and above, is acting to slow relative spin rate against the direction of motion by introducing a reverse spin. The angular inclination of this circular movement is referred to as "tilt". Any surface now viewed along the electron shell will appear to have a circular motion introduced. It is natural, occurring in a top as it slows and first begins to waver. Although not readily apparent, depending on electron orbital speeds, this operation would be happening at very near light speeds and could actually have a very strong effect on spin if "tilt" angle was large. The actual tilt angle has not yet been recorded and may be slightly different for different elements or atomic structures. It may be different for ferromagnetic elements which have fewer electron orbits intersecting the magnetic polar regions of the atom, but is more then likely related to their atomic weight. Heavier atoms having a different overall resultant tilt, thus radiate a different gravity field.

I find this to be a somewhat credible partial theory of gravity, because devices now exist which loose a large percentage of weight based on this principle discovered 100 years ago when emulated in very heavy fast spinning wheels being struck at the center of axis then lifted with one hand from the end of the shaft. With this theory, Gravity has added one more element of "spin" to the spin of time, and modulates the electric potentials of the atom to radiate outwards in a moving 2D field. The earth is on a "precession" know as the "platonic year" about 25,800 years long.

View from our simple model:

As the Gravaton pairs spin inside the electron shell, we see the shell in operation as one cohesive unit with all electrons spread out as far as they can get from one another. Therefore as the higher frequency forces of gravity circle inside the electrons they produce a vibration seen in the entire electron shell. This leads to the movement of "precession" or "tilt" of the whole shell. The movement is explained within the gravaton pairs amplitude of vibration.

Induction of Gravity

View from outside sources:

If gravity is an area type "field" with an electric component, it may have both attracting and repelling fields of force in some form. Like electric potential repelling other like forces away, the "attractive" force of gravity may actually be an induced reversed field set up in nearby matter through the movement or shifting qualities of the field. It has been speculated that gravity may be the only electric like force that radiates outward from an atom to any great distance, as most all others are nearly balanced within it. Moving further away, as you increase distance outwards from it. Less is cutting through the matter it encounters, but higher voltages are induced overall, as observed in the atmosphere of the earth increasing its local + static charge as you move higher.

View from our simple model:

Gravity is seen to have both attraction Gravaton, and repelling AntiGravaton, spinning with the reference of [Source], and changing between dominant flows based on the time spent of each particle in this density the "wink on time". As the helix of electrons traverse a tight curve encountered within the atom, a high magnitude force is created by unbalancing the two flows at its point of highest magnitude. The vibrational spiral within its orbit around the atom.

Spirals

There is a "rifling" effect present with any projectile in our local space. A clockwise movement, viewed from behind, as an object is propelled at high speed. Guns are now built around this effect. The electron may also have this movement as it moves around the atom. This may be the source of a field which keeps the electrons from smashing into the nucleus. A repelling force that has been calculated to originate at the electron shell of atoms found in "electro gravity theory," which may be the source of gravity. Picture an electron moving in the pattern of a spiral path as it moves around the atom at near light speeds. It becomes a very tiny magnet with a spherical torus field. Because of its speed it possesses the qualities of high spin and exists in a different time frame.

As we are now viewing the two spirals of Aether wrapped around one another emerging from source as a **[helix]** this is interesting. Now the overall winning force of gravity is dominant, the Gravaton inflow and may reveal the direction of its spin in the helix of the Proton. The Gravatons inward flowing component may now be viewed as a spiral rotating clockwise from behind as it is sucked into source. The AntiGravatons motion may be viewed the same only direction of Aether Flow has been reversed, and now rotation polarity has been identified. Antigravatons will be shooting energy into their field in a counter clockwise motion as seen from where they emerge, and Gravatons will be sucking energy inwards in a clockwise motion for the Proton. The energy from [Source] flowing both directions forms a helix pattern.

Kosol and Koeun Noun Ouch Spherical Generator

Important note:

An electron that is not in motion or that is in straight motion may still be crossing the density layer but will not radiate a dominant gravity. As it encounters a curve, its helix will begin to release either more outflow or more inflow, resulting in the predominant Gravity polarity resulting. We may now consider that as the Electron has a negative charge it may be possible for this to occur with **Antigravity being the result**. The assumption is now that by altering the overall rate of spin for the whole system, its torsion field, we can cause the opposite of gravity, just as with voltage. All we need to do is mimic the Electron with a much different curve.

Placing Protons in motion may be the difference, actually spinning the device will effect both the [Protons] paired gravatons [distance from center] and the Electrons oppositely.

Gravity Control

It is obvious that through the use of magnetic fields, ESR, and NMR frequencies we can manipulate the motion of atomic particles, changing their natural frequencies of precession and flipping their magnetic polarities. How we can get them to supply free energy or anti gravity is the present challenge. If gravity is a resonant coupling of atomic spin, then possibly by flipping enough nucleons to a high energy state then tilting them 90 degrees to electron spin may decouple the outflow and inflow of gravity for a time. In this state gravity would no longer be a spherical force but a bidirectional force like a magnet. All atoms spinning coherently. Two of the proper atomic resonant frequencies crossing at 90 degrees may tend to accomplish this. The cancelling [bifield] winding at 90 degrees to the main field winding may be doing something like this. Neutrons may align with a bifield cancelling winding and electrons with the main one. As we energize them both atomic spins tilt opposite directions.

In field theory, gravity control becomes, either suppressing the "induction" of gravity within an object, generating a reverse field, or altering one or more fields thought to effect gravity, torsion, electric, precession of spin, or pushing the atoms towards a density threshold. Exciting atoms and raising their frequency, lowering their distance from source. Flipping their magnetic polarity at the proper time, or rotating their axis of spin to produce torsion. Any method of aligning the atoms axis to [lock in] or alter [tilt] for a time, may produce weightlessness by removing the atoms ability to respond to gravity, decoupling it.

Sound vibration, electric field rotation, energy flooding, reverse precession, deep meditation, magic...... Strong magnetic fields produced by super conductors. Spinning pyramids. Creating a large torus field in which these parameter may be controlled in a localized area of space. Searl disk, Kosol sphere...etc. Van de Graph generators do not lose weight. However a large static gradient may help to push the atoms towards ZPE threshold. With superconductors a very strong magnetic field has been claimed to block gravity partially. As with Hamels spinning cones, magnetic fields may possess qualities to block or redirect gravity.

Gravity is affected little by a potential charge, but it is affected by its curved path through space.

Gravity control may be the amount we can alter the helix pattern of the [gravaton and antigravaton] within the electron the neutron and the proton as they spin around center at the point of threshold where it phases into our density. Using the electrons dominant counter clockwise entry and the protons dominant clockwise exit we may be able to engineer devices to bend electron and proton movements to alter Gravity either way.

The forms of matter:

Solid, Liquid, Gas, Plasma, Aether

Forces Compared

High magnitude forces:

A quadrature relationship is observed between torsion and gravity. These forces are linked and may be source of the curved space time fields found in conventional Einstein theory. Rather then space being curved, it may be torsion or time which can be viewed as curved. Time moving slower causing a strong gravity field.

Low magnitude forces:

A quadrature relationship is observed between electric potential and magnetic field. These forces are linked.

Comparison:

The main difference between these forces may be merely the frequency of rotation generating them, or the distance from center of spin. This leads to the realization that forces may be separated into different bands of operation, interacting only with other forces of nearly the same wavelength through resonance of vibration. Thus a magnet weighs the same no matter which way you set it. Gravity is acting on its smallest movements of matter and magnetism on a larger movement.

Electric force corresponds to gravity, it is an inflexible force, unforgiving and ever present in space, falling off as a distance squared force. Magnetism corresponding to torsion or the curve, an elastic force originating with spin, allowing for compression and expansion. Although torsion is presented here as a 1D force of the gravity torsion field, and magnetism a 3D force, the same model can be used to view both sets of forces. A circle moving perpendicular to its spin axis. The north south poles of a magnet can be viewed as a linear model, as well as torsion being viewed as a volume that contains the gravity component.

Torsion seems the most flexible of the forces and is shown to be alterable in matter that is spinning at high rates for a period of time. The torsion component lags the motion changes, but is slowly equalized over time.

Since magnetism contains a torsion component it is a flexible force, magnets floating can be easily compressed on one another and will bounce to stability for a period of time. Gravity and electric force on the other hand do not seem to offer this flexibility. In electronics oscillations result only from moving electrons energy into a magnetic field and back to an electric field as in an oscillator. Current lags voltage in an inductor, and leads it in a capacitor. The electric force setting in a capacitor offers little flexibility, and is not seen to bounce in any way, neither is gravity seen to bounce or show natural flexibility.

It may be that the only way to affect gravity is to work with the torsion field, or altering spin. Just as the energy in electricity is seen to be moved between the electric and magnetic fields, the energy between gravity and torsion may work the same. The tensor or scalar coil may show us the key principle involved.

Theory of Gravity and Aether

Stepping out of the Theories of Fields and into the Aether brings up several paradoxes as well as magical intuitive qualities. These must be resolved to reach comprehension.

Aether:

Aether is everywhere. "Intuitively" it is necessary to connect force over distance.

Aether does not appear visible in the half of the universe we see, but in the underlying opposite one, intricately coupled to all fields, forces, and even space through magic. "Magic" being the quality things possesses which we can not "intuitively" comprehend within our present state of awareness, but would use to describe known reality perceived. Aether has a tendency to become a "catch all" for all forces we can not explain yet do possess an intuitive awareness of. It has been compared with Chi, Ki, Love and many other of the mystical forces found in "intuitive" systems of healing and Religion.

Stationary Aether: Aether [1]

Aether may be viewed as the fabric of space.
Transferring waves at light speeds.
Transferring or storing spin elastically.
Holding temporary spin or torsion memory.
Transferring gravity as a surface or 2 dimensional force.
Transferring magnetism as a 3 D force and electric potential as a 2 D force.
Maintaining momentum as a linear force.
Creating centripetal and centrifugal force.
Becoming a strong wall at near light speed.
ZPE being the sum of all waves moving through it.

Viewed this way Aether becomes the magical interface between an invisible universe and this one, sharing only light speed as the one constant.

Basically the Gravity - Torsion fields transfer medium.

Movable Aether: Aether [2]

Aether may also be viewed as a substance that may be moved, warping the very 3D nature of space along with it. Holding a local background pressure or flow know as ZPE or zero point energy. Flowing up and down as gravity, flowing in and out of magnets, being pressurized as voltage. Circulating through our bodies as Chi. Any invisible force, even the wind, which flows. Imparting the "intuitive" existence of one underlying source for all these manifestations. [Oneness of God]. This is where things may become somewhat paradoxical however. Because if the Aether can be moved, what exactly is it moving through? Aether [2] must be moving through Aether [1]. The first premise of Aether [1] was to explain force over distance. If in fact Aether [1] can be spun, or small quantities of it sucked into vortexes as with tornadoes, manifesting on this side of the universe as atoms. Then it is hard to explain how fields may still move through it or how it can move through itself. The same would hold true for magnetism and gravity, being labeled Aether [3] and Aether [4]. If they were the same Aether flowing, then why do they not affect one another?

As Aether [1] moves, light speed would change. At points of high spin in the Aether [1] waves would not travel through at a constant velocity. So indeed this may be true. If Aether [2] density can be decreased, sucking all energy out of an inflowing area into a sucking vortex, then gravity may be simply a low pressure area of Aether. Magnetism a 3 dimensional flow of Aether [3], and electric potential as well as gravity an area pressure of Aether [4]. Time may still be the local spin rate, however it would not be expected to sum vectorally anymore as with fields, but instead be a point location effect different for every point in space dependant on Aether[1] spinning rate at that point.

Extrapolating Aether [2] that moves through Aether [1] is a complicated idea, and may be worth perusing. Keep in mind this is a very easy way to loose track of differences in forces and create confusion. Merely seeking to redefine all the known forces as Aether[x] may become redundant? It is my contention that if we want to explain invisible forces as a flow or a pressure we simply use their proper names. IE Gravity flows up or down, magnetism flows in or out, Chi flows through the heart charka, and leave Aether to its original Aether [1] meaning allowing a fabric for all these flows to move through. Due to the obscure nature of Aether [2], field theory seems to offer more tangible definitions when adding only Aether [1].

Aether Gravity:

Whether gravity is viewed as a wave moving through Aether [1] or a pressure of Aether [4] flowing is an interesting question. Both models, "field waves inducing attraction," as previously presented, and "Gravity pressure" should be explored.

Gravity as pressure:

If gravity is a flow moving from high pressure space inward to low pressure earth then several qualities should be present. For a static pressure gradient to be present then there must be a constant sucking in inside matter and a constant replenishable supply coming in from space. If this is a quality of gravity flow then its pressure should drop as a "volume" moving towards the earth. It should be an inverse distance cubed force. It is this glaring inconsistency distance ^2 verses distance ^3 quality of gravity that makes me want to dismiss the Gravity pressure theory and return to field theory or waves moving through Aether[1].

Gravity may also be viewed as two particles in opposition as they enter our density, spinning two possible directions in Proton and Electron. One is pressing outwards and one is sucking inwards. The Aether will flow through either one as it crosses the threshold of our density, bringing in or out either flow. Regulating the helix motion of the two particles as they move through space 90 degrees to their rotation in a curve may regulate which one dominates and spends more time at threshold. So the Aether theory has explained many of the observed qualities of Gravity, as is found at the core of its entry into our density.

Gravity as background or ZPE waves in the Aether:

It has also been presented that Gravity may be a result of waves that move through Aether [1] consistently in all direction pushing on matter in all direction of a bandwidth of wavelengths capable to achieve this magical quality. In essence background invisible light pressure. As we stand in the shadow of other matter a percent is absorbed by that matter causing a reduction of outside pressure from that direction. Does this model follow the inverse distance squared quality of Gravity? This would seem to be the case; you stand in the 2 D shadow of an object so as you approach it the shadow would follow an inverse distance squared increase. The refuting evidence I could offer would be these things.

The patterns of energy in the galaxy tend to be radiation coming from its center and are not constant. Sun spots would affect small changes in Gravity.

How would introducing precession into rotating iron wheels cause weightlessness? How could vibration produce weightlessness? Why does gravity increase in deep caverns as opposed to on the earth's surface? None of which seem to be explainable in this model.

The Earth and gravity:

Gravity has been shown to be stronger on the oceans and even stronger in deep caves then on high mountains. It continues to increase as we move in. How far has not been measured, but we are aware of two different layers of changing density within the earths core. We know that a large enough body of matter will achieve gravity strong enough to break apart or crash the electron orbits, creating a density packed neutron star. Thus a hollow core is very unlikely. This is evident with the ability of gravity to extend itself by the distance squared function rather then falling off as a cubed function like magnetism, being able to reach out much farther then simple pressure in 3D space or a magnetic field. The earth has a predominately positive electric field. As you move higher in the air the + charge becomes higher rather then lower seeming to defy electric laws until the outer ion electron belt is reached. Gravity may be causing this through induction of movement from its center, where it can be seen that movement is greater moving outwards on a wheel or a spherical field in motion. The "electric" quality of gravity may be inducing potential in the air.

The Van Allen belts show us that it is possible to have protons orbiting a magnetic field as well as electrons. The proton belt is located inside the electron belt, and both exist at quite a distance from the earths crust both following the curve of the earths magnetic field where it is really very weak.

Aether Time:

Spin or multiple spins may be present in the Aether for any local volume of space.

Time may be the relative background Aether spin rate rather then its still quality.

The Zero point of local space may be the background Aether spin relative to light speed.

Altering the spin of Aether may effect time, as with torsion fields.

Spin in Aether can be altered and its effects last for a period of equalization time.

At high rate of speeds Aether [1] becomes elastic and alters timescale.

While it is possible that altering the density of Aether may effect gravity, altering the spin of Aether may effect time.

Spin may be thought of as a movement of Aether, or a movement of us through Aether.

If Aether spin is the rate of time flow then tapping the ZPE to extract energy may alter time in a region of space.

Time being the incrementing of spin [a constant] in the density you are located will vary, based on your DS factor [distance from center] which effects your relative speed of awareness to light speed [AW factor].

Aether moving through 3d space as a gravaton pair has torsion, and thus creates a field where time can be observed.

Vibrational gravity effects:

It was noted that experiments with sound have shown it may be possible to make weightless a piece of uniform material like limestone with the bombardment of 5 frequencies of high volume sound. Using salt as an indicator, each low frequency was found through trail and error. Certain frequencies are seen to form patterns in the salt as it vibrates on the top of the material. By applying all 5 at once the material became weightless. The vibrational effects are unknown at present as to how this may operate but noteworthy is the 5 sided nature of the platonic solids. And that the resultant flux operation may also be viewed as a vibration. Keeley stated that only three frequencies were necessary but did not elaborate. If atomic resonance was a factor then maybe we just need to explore how vibrations c an be used to alter the "precession" set up in a spinning object to solve this puzzle.

Cold Energy

Our present model of electronics has been built around observing the nature of Electrons. Electrons always seem to display [expanding] qualities. Even a positive potential has been described as merely a shortage of electrons around atoms carrying protons with less electron balance. They are referred to as electron holes. Electric energy is propagated along the electron orbits of atoms which are loosely bound. The Sweet VTA was a device that seemed to tap a new much deeper source of [cold] or negative energy. Coming from the introduction of Barium atoms into a magnet, the cold energy seems to have the qualities of absorbing heat. Also the cold energy is theorized to be capable of propagation through very small wires and even powering devices. The Sweet VTA showed a tremendous power gain over conventional devices, as well as a loss of weight. Negative energy however instead of electrical burns may cause frostbite if shorted through the body. Literally a freeze flow, absorbing energy wherever it encountered it, manifesting the opposite qualities we observe with electricity.

The device Sweet built resembles a coil found in Smiths work instead of an iron core it uses a barium magnet wrapped on all six sides with three coils. All three coils are used during magnet programming, and only two are used to tap power. By conventional electron theory this coil arrangement appears to do nothing. However as the pattern keeps popping up it deserves much more investigation.

As normal electrons flow between atoms, hopping along, electricity is propagated down a wire. The wire is a [conductor] with electrons in loose orbits where energy can be added then removed. With the [cold] electric flow, energy is conducted possibly down an [insulator] or a barium laced alloy. An insulator has the qualities of having its electrons stuck very hard in their outer valence orbits. There are no free electrons to flow energy. What ever is flowing down the cold wire is transferring, not expanding energy, but contracting energy. This makes sense if you consider electric flow from the standpoint of waves traveling between atoms rather then electron jumps. The negative energy present at the nucleus of the atom, having contracting properties, such as with the proton, may propagate this energy between atoms, through certain atoms like barium, that may be viewed as having loose protons, or tight electrons. The protons absorb negative energy then release it much as electrons do with positive energy. The same propagation would be expected to work, and this may explain some about dielectric constants that are always present when normal electricity flows in a conductor through a wire with insulation around it.

Also however it is noteworthy that the Searl device, Helsinki, and David Hamels cones produced effects that opened the inflow of nature at some point. Drawing inward large quantities of cold energy, exposing our world to the direct drawing in of energy to source.

We have already identified the Proton as the overall source of negative energy or inflow with respect to gravity and electronics. An energy propagating between atoms based on this type of energy may more properly be viewed as [Protonic energy]. This now gives us the whole picture as to the spectrum of both yin and yang aspects of energy flow, and explains why in electronics two positive voltages will still repel. They are merely different levels of the Electrons expanding energy.

Protonics:

The study of how protons propagate negative energy potential. As in our observations on the helix model we identified that both the electron and the proton have access to inflowing and outflowing forces. Due to the bends in their helix they manifest in two different ways. The electron resultant force is expanding outwards, and the Protons is compressing inwards. Thus Gravity, an inflow, is possible because the proton is winning in the fine balance between the two forces. However in electronics the expanding qualities are ever present, as electrons will even jump or arc to get away from one another. Protons then may be expected to draw together with a contracting force. Thus design of propagation devices must be altered to accommodate this type of energy. The effect of weight loss in protonic flows is also noteworthy, and indicates that at the proton it is possible to divert the gravity effect into a flow of energy.

On the one hand it would not be expected that proton energy [protricity?] would want to jump off wires as it has a compressing quality, however we still would have no desire to receive freeze burns from contact with it. Some form of protonic insulator is needed for a wire covering.

Noteworthy is the fact that barium is presently used in automotive spark plugs.

A study of the perodic table of elements shows us the electron conductors on the center to right of the table.

The semiconductors a little further right. The Protonic conductors may be the ones to the left of the chart around barium. Calcium and Niobium present a interesting presence, as niobium is available in small quantities in arc-welding rods and calcium is found in some quartz crystals as well as in lead calcium batteries as an electrolyte. These elements would be seen to have much tighter electron orbits and may be good conductors of protonic energy.

Conversion of electronic to protonic energy:

The coils found in Sweets work as well as Smiths magnetic work are observed to have one coil that is not apparently functional. It is wound at right angles to the other standard coils and has a bipolar canceling nature.

With Smiths coil there is a normal coil wound on the center of a square figure 8 core. This causes both sides of the magnetic field to be split evenly between the windings of the 90 degree coil around the outside of the device. This would seem to have a canceling effect on its output if it were wound with conventional wire. In the Sweet VTA the two normal coils sit to the side of a permanent magnet that has been specially imprinted with a frequency as well as another normal electrical current. The ninety degree coil is bifield wound and seems to cancel out the normal electric field.

It is not evident how the VTA transfers energy to the light bulbs connected however they do seem to light just fine. It is possible that the process converts the protonic energy back to normal electricity or that the insulation of the wires is the transfer medium. This is unknown.

It may be interesting to replace the output winding of the Smith device with [protonic wire] and see what the result is. Study may be done in this area which may open us to the ability to channel this negative energy originating in the nucleus of the atom.

Conclusions:

If protonic energy turns out to be a viable working application, then the whole theory of electronics may be altered to explain both halves of it. [Electron Proton Fields] EPF. Discovering the connection between magnetism as generated between these two flows may be the crossing point. If they do lay at 90 degrees to one another the solution may be simpler then we could imagine. If other more complex energy transfer is necessary we have only to discover them. The side effects would be protonic devices that [**over cool**], rather then [**over heat**] as with electronic devices. A device using both methods would reach an efficiency much higher then in conventional electronics applications because both heating and cooling if allowed to interact would keep components working at normal temperatures and could be stressed much further.

Planar Angle of Spin

Looking at a circle which is the result of spin of an atomic particle such as the Electron and the fact that it forms a precessing magnetic field, you can see that a planar surface may be extended off the circle of motion that extends as far as one cares to follow. Many spin planes would exist for spinning particles, however they all may lie parallel to one another. Due to the angle they sit at in the universe on their smallest rotational axis, that of particle rotation or angular momentum.

If Wilbert Smiths model of spin is an accurate representation then it is a given that at some level of spin in our model of the universe, spin must be constant across all matter in a local region of space. For matter to be cohesive and have tensile strength on some level it must spin in sync. Planks constant lends us to believe that the spin of an electron [its angular momentum] is one of these constants.

It is possible that electrons all spin in the same general plane of motion, not in their orbital movement, but in there self spin. All protons and all neutrons may as well have particular angles of spin that are constant across all matter. This may be the spin linking gravity and time across our density.

As orbital particle motion is not constant and can be altered with NMR, it is probably not the root source of Smiths "Spin". Whatever Smiths "Spin" is that must remain constant, the claim is that altering it will cause time and gravity to be altered, and if it is altered far enough in too small a space then matter will fly apart. If the frequency is raised we will be pushing mater into the higher densities.

If the angle of these spin planes can be detected, then a directional element may be present across the galaxy that can be used as a form of directional indicator, much like a compass on the earth. If the spin planes are slightly different for Protons, Electrons, and Neutrons, then a 3 D direction finder may be present across all matter everywhere. Up and down could be given labels constant across the universe. Also it may be interesting to do rotating experiments in these spin planes, to see what is possible when spinning atoms are aligned with the angular spin of atomic particles. Aligning a spinning experiment with the earth's magnetic field or its axis of spin may not be as silly as it sounds. Just because the Earths magnetic field is very weak, it may still be pointing to something larger within mater.

Also the spin plane formed by the earths orbit around the sun may fall into a "universal spin plane" for all the atoms gravity components as well. Aligning an experiment with this planar angle of earths orbit around the sun could lead to a discovery of atomic proportion!

Measuring the spin plane angles may be possible. Sensitive meditators have been able to sense directions on the Earth for a long time. Possibly these spin planes can be discovered as well, measured and given universal charted angles.

Occult Chemistry

Originated by Leadbeater, and developed into an almost complete periodic table today, Occult Chemistry is worthy of study.

For each element different structures are perceived in the higher planes of existence.
Some of my personal intuitions gained from this study follow.

Intuitions from the Occult **Anu** [The smallest quantum structure of Occult Chemistry which appears in both male and female versions]. 3 forms of [electricity] were perceived in the structure of the Anu as three strands.

I offer this parallel:

1- electronic, heat microwave energy
2- protonic, cold [electricity]
3- neutron or scalar [electricity] opposing fields which reach out very far, at high speeds, but are not measurable as electric force.

The three forms are perceived under the heading of [electricity], which is really only [electron flow], but was the only [mental symbol] familiar to the seeier for comparison at the time the system was originated.

Kosol and Koeun Noun Ouch Spherical Generator

For the seven colors perceived as seven strands I offer this parallel:

The infinite frequency of ESR and NMR seen as perceivable to us through the seven channels of our reality, and the mental symbol of our chakras and our color spectrum perception. This is however a function of human perception, and our senses do not allow us to see the full spectrum, which is also present in all the forces. Perceived only as seven bands of frequency, yet infinite in actual frequency range. To me this is a representation that within the Anu is the capability of the **three forces to vibrate in all the perceivable ways**.

The Anu was seen as a heart shaped mobias strip of 10 strands. Twists at the bottom of a helix center and a spherical heart shaped outer layer. I would offer that this psychic vision of the root structure of creation would reflect a meaning of ultimate Love to the seeier and thus may appear as a heart shape for this reason.

There may be much deeper meanings within this system, however I would suggest treating it as though it were a psychic vision rather then a measurable model, because that is really what it is. As of yet I have found little meaning in the numbers offered, yet a very accurate perception of what would have to be present in a structure that could offer creation all that we find in nature. The fact that the smallest curls perceived look similar to current drawing of magnetic flux is amazing considering when this system was developed.

Section Two [Devices]

The SEARL disk, the David Hamel cones, the Helsinki experiment, as well as the Kosol device, point the way to setting up a series of magnetic interactions that combine fields which may both attract and oppose one another in a pattern that may cause the Elastic qualities of the Aether to start to move or vibrate in the pattern of a "spherical torus" centered around them. The torus, although magnetic in 3D structure, may be electrical in nature having a gravity field appearing around it and possibly a time displacement "torsion" field as well. The field will affect all fields within its volume, including "relative" momentum, gravity, and time. With the creation of a spherical torus of this nature, the engine becomes the craft as well.

This work is dedicated to the "comprehension" of such devices, and methods of control.

Scalar Canceling or Tensor Devices

The use of a [tensor] or opposing magnetic compression force may be a necessary element of ZPE and density crossing devices. The simplest of these devices, two magnets glued together in opposition and a coil wound 90 deg to the intersection plane, where reverse magnetic fields come together. This technology offers a possible wave propagation through time. This could lead to cross density communications as well as energy inflow or outflow control directly from source, as atoms are powered. **As scalar canceling technology is not yet mathematically understood these devices will come only from**

Kosol and Koeun Noun Ouch Spherical Generator

experiment. My own butterfly coil is shown to produce a long trough of tensor area that can be used for experimenting. The **circle antenna** may also prove effective in cross density work as it offers a larger area of tensor space to work in at a higher frequency. A frequency of 2 times the diameter will create an area of tensor stress inside the loop. Interesting to note this is not the normal resonant frequency for a loop of wire. The common scalar coils being wound as the Smith coil offers only very small areas and I would think communication would be the most you could hope for. Time and experimenting will bring more of the pertinent questions.

Smith coil:

The original coil described by Wilbert Smith is simply a coil wound with each wire loop in opposition to the one above and below it. As electrons pass opposite directions through each winding the magnetic fields oppose and cancel the EM band. The only force left is one of torsion, or as Smith refers to it the **tempic** field. So the coil was intended to **remove all magnetic and electric components** from radiating and leave only a shell of altered tensor energy, or altered time flow rate emitting from it. The true Caduceus coil does not have wires crossing over one another, as you can see from inspecting the symbol denoted as two snakes coiled around one another, mouths facing across the top.

On a ferrite core, the Smith coil uses one wire up the coil then back down, if wound at the same time from the center of the wire, they cross over one another on two sides of the coil form.

The addition of spacing the coils such that wires cross at 90 degree angels adds another component to the coil. It allows only half cancelling, the rest of the signal is added and crosses the canceled field at 90 degrees similar to the magnet description above. Thus one coil is able to create a [carrier tensor field] and modulate it with EM at 90 degrees.

The tensor beam can be thought of as a laser beam that does not spread out as it propagates, it is an altered torsion force and travels linearly as a torsion field, or as a cylinder of altered torsion, and thus may alter gravity. This would appear to combine the 4 forces in one device, using now a torsion carrier modulated by an EM field. Once again it is [time] seen as the flexible force, rather then gravity, being modulated by this coil.

Another perspective is simply, a magnetic field is a minor alteration of the time flow rate. Canceling the electro magnetic component frequency band between two opposing magnets, still leaves the altered time rate in tact. The time field propagates at superluminal speeds exceeding that of light speed. As photons move "through" the time field. This coil structure was said to allow communications across great distance. The downside is that it must be perfectly aligned, or it must be spun as with an atomic particle, because its field does not spread out. Torsion is a 1D field.

When this coil is stimulated with Microwave frequencies, it may actually go into electron magnetic resonance ESR. Coils stimulated in this fashion have been seen to jump around physically.

Tom Beardens Scalar Potential and the Vacuum

Beardens Scalar Waves:

The theory on Scalar waves is rather remarkable and worth investigation. The nearest I can visualize are waves generated by opposing magnetic fields. The magnetic fields cancel, and pressure or [Scalar] waves are emitted from the junction when a third magnetic field is pulsed at 90 degrees to the opposing fields. Tests with a scalar generator built from 4 magnets on a flat magnetic bar with a driver coil have produced pulses that stimulate the nervous system and feel like electricity is being beamed into your nerves. The beams are very straight off the ends of the singular magnetic field. I have observed however that these waves do not travel through everything, it seems a magnet will stop them.

The basic model is 2 opposing fields held together creating a tripole magnet. The third magnetic field intersects the blotch wall at the center of the opposing fields where tension is highest. Causing either of the crossing fields to vary or pulse will generate scalar waves. The atoms at the blotchwall will tip up to 90 degrees away from one another forming perfect 180 out of phase alignment as they absorb photons. When the field is returned the atoms spring back at an ESR frequency and emit photons that are also 180 degrees out of phase. The waves will be in the microwave band and depend on the magnets field strength as with ESR.

This technology is very useful for determining magnet placements on the Kosol spheres. By placing magnets on inner and outer spheres with an attracting field, opposing magnets on the middle sphere may be set 90 degrees off with like poles together, generating scalar waves as the spheres turn. More research is under way to determine if this is the mechanism that may cause rotation in the Searl disk.

With Beardens model of vacuum energy as [Source], scalar waves emitting from the center of the universe may be powering everything. Learning to unbalance these primal balanced waves may be the key to unlimited energy. Using interference patterns may be the key to conversion.

Legend of the Circle Antenna

[The disappearing paper clip]

This was a very noteworthy report I read from a fellow who had an interesting experience as a child. He was playing with an old UHF antenna and an automotive ignition light circuit board that pulsed high voltage spikes that lit an intense, probably neon light, when it sensed a spark from a pickup coil placed around the ignition wires of an engine. He wired the circuit board output to the looped wire UHF antenna, placed a paper clip at the center of the circular antenna hanging from a string tied to the top of the wire loop. He hit the antenna with pulses from the circuit and the paper clip slowly faded then disappeared. The string was still hanging as though it held the paper clip. He then pulsed the device again and the paper clip reappeared.

This memory stuck with him for life and his account was found on a site years later. He had no explanation and often wondered what really happened.

Seeing this from the eyes of a scalar or tensor interaction it becomes explainable.

By chance, the frequency of pulse hitting the circular antenna was an exact half wave of the wire loops diameter. This produced a total canceling effect of the EMF at the center of the antenna leaving only the torsional component. The time shift caused the paper clip to shift density and fade out as with the Parr wheel. A frequency of around 430 Mhz on an 11 inch loop should produce a similar effect. However since the pulses were not sine waves possibly a Tesla type [short pulse] was involved as well. This experiment deserves further investigation. Since a loop antenna resonates as a full wavelength antenna, the diameter is circumference / pi, then the wavelength = (circumference/ 2 pi). Thus every point on the loop is exactly 1/2 wavelength from its opposite side of the ring. This relationship should produce a ring of cancelling EM waves leaving only the torsional [tensor] stress as a radiated field. The circle will intensify the effect at the center crossing point of all waves around the ring where all torsion will intersect and all EM fields will be canceled.

Kosol and Koeun Noun Ouch Spherical Generator

The Delta T Antenna

Placing two loop or quad antennas at 90 degrees to one another will produce a rotating magnetic field if both are fed 90 degrees out of phase. Much like a 4 pole induction motor working at 60 cycles, the magnetic field will rotate equal to the frequency hitting the motor. A compass held at center of the two antennas will spin if the field is kept slow enough, showing that the magnetic poles really do spin.

The claim is that when the frequency of such a device reaches up to around 300 Mhz the speed of the rotating magnetic field begins to exceed the speed of light at a distance close to the device. Moving outwards from center where the wavefronts 90 degree magnetic vector of the moving magnetic field hits light speed, is expected a corona to form around the device, probably a spherical boundary. At this demarc point would be expected one of three things. Waves vibrating faster then light cross the density threshold and disappear from our density, or containment of photon energy is compressed along this surface creating a force field not able to exceed light speed, or waves would merely slow down and appear out of phase to incident waves causing a loss of resonance and a cancelling out of off phase energy.

velocity of wave = 299792458 [meters/s]
wavelength [meters] = 299792458 [meters/sec] / frequency [Cycles/Sec]

An antenna that has a 1 meter diameter loop, has a resonant wavelength of 3.14 meters in the X axis. This gives it a resonant frequency of 95.475 Mhz.

Distance around the delta T antenna in the YZ plane is 3.14 meters as well which the magnetic wave will traverse spinning in one cycle. The magnetic field from this antenna will be at the speed of light just at the outside of the YZ plane diameter of the antenna as with any resonant curved quadrature designed antenna at resonance.

Practical experimentation with such antennas shows that the wavefront is really slower then light speed inside the conductors in the X plane, reaching up to .95 the speed of light, and not the theoretical limit of 299792458 meters/second but more like 284,000,000 meters per second. Placing the light speed limit outside the antennas one meter diameter to about 1.05 meters.

About 5 cm outside the 1 meter Delta T is an invisible light speed limiting sphere, where a magnetic field is moving through any objects at that point faster then light and **may begin to effect superluminal vibrations like gravity and time**.

It is unclear why 300Mhz was offered as the lower limit, as any Delta T antenna should be the same at resonance. A frequency of 300 Mhz makes the antenna diameter about .318 meters, for a circumference of 1 meter wire loops setting at 90 degrees. This would tend to place more energy towards the center of the system, and light speed threshold about 1 cm out from the wire loops of the system in the YZ plane. Higher frequencies could be used by hitting the antenna with harmonics of its wavelength bringing the interactions closer to the physical antenna, but they will never exceed light speed on the wire, or inside the loops because RF does not exceed light speeds in the conductors of an antenna system.

It may be interesting to place one antenna system inside another and observe how they may interact at different resonant frequencies. Placing the outer one exactly on the threshold point of the inner one, or even spinning it the opposite direction. Also using the pyramid shaped design with a frequency of 2x may produce a field that is more coherent top to bottom.

Kosol and Koeun Noun Ouch Spherical Generator

The Three Fields of AG

The SEARL device, the Helsinki device, and the MAGVID share a common pattern. The creation of a spherical energy field surrounding them. At Helsinki called a blue plasma, SEARL called it a protective force field or shield, and the MAGVID calls it an electron shell. The shell seems to be the product of three separate fields at work which become apparent as you study the devices. The formation of the shell may preceed the AG effects, but after the AG and ZPE effects the blue glow was definitely observed. The energy present finally reaching up into the light emitting frequencies where it becomes visible.

The first field [magnetic]:

The deflection field. An overall magnetic field surrounding the device, mirroring the earth's magnetic field. This field deflects electrons outwards at 90 degrees to its magnetic field flux lines and pushes them into a circle around the magnet very close to it. This is a natural interaction of electrons passing a magnet, they are deflected or propelled at 90 degrees to the magnets flux path. This field forces electrons away from the core using a distance cubed force, operating very close to the magnet generating the field. It also pulls them back to center as they pass the magnet, moving them into a curve around the device perpendicular to lines of flux encountered. This field is also the primary filed that opposing magnets at 90 degrees can produce scalar waves along.

The second field [electric]:

The containment field. An overall positively charged center or inner cylinder area. This static electric field pulls electrons towards center and keeps them from escaping the system if they wander outwards. It operates as a distance squared force and reaches out further then the magnetic field.

The third field [scalar]:

The excitation field, usually opposing magnets pulsing at 90 degrees to the deflection field. This field activates the tensor areas where electrons are set into motion with velocity which can escape their small magnetic fields as they are setting within the negative area of the containment field. The tensor area is where energy is flowing into the system cross density. Source of cold energy, this field may be producing a torsional or time alteration force as well in the entire electron shell.

The combination:

The combination of these fields creates the electron shell or plasma where electrons are constantly being drawn in and can not get out which tends to build over time. The excitation field accelerates them towards their normal atomic speeds or vibrations. The manipulation of this electron layers vibrations by the tensor energy alters torsion and thus gravity for the whole system. If the combination pushes towards a higher density then gravity is reduced and time speeds up for all atoms effected. It is suggested that this electron layer operates similar to an atoms shell, and may work across a very large spectrum of frequencies based on the precession present in the shell as in ESR.

As to spin direction and multiple spheres:

It should be noted that when electrons approach a magnetic field they do not turn the same direction when approaching the outer flux path along the outsides, as shooting outwards through the same flux path encountered inside the magnetic field.

The direction reverses from inside to outside as the flux path curves around 180 degrees to where it leaves the top of the magnet and moves down the sides. Spin direction is the same top and bottom outside and it is the opposite on the inside of the flux field. This effect must be considered when trying to form a shell of electrons moving coherently around the device and especially

Kosol and Koeun Noun Ouch Spherical Generator

if it is expected to use electrons to accelerate the shell of a metal device. Using the rules of magnetism, if we carefully consider which directions electrons entering the field will be deflected as they approach flux lines from all directions it will become obvious that they will all be turned the same general direction as they encounter the field and spin clockwise looking down at the north pole along the entire outer flux path. If any electrons penetrate the first field and hit a second one from inside they will be reversed and work against rotation. The magnetic field must be strong enough to turn them all, or it must be routed through a metal sphere around the magnet so that none can reach the cores primary field from a secondary outwards angle at high velocity. Also it should become clear that in the case of a metal sphere conducting a central magnet flux, electrons will bend opposite directions when approaching the inside of the sphere or the outside. Thus electrons must be kept from shooting outwards in the air. However alternate spheres could only be expected to rotate opposite directions if they were polarized with opposite or aiding magnetic fields. It would seem the only way to make multiple spheres spin opposite directions is to magnetize the spheres themselves with reversed fields top to bottom, as one overall field would cause all to rotate the same direction as ZPE is entered. Thus we see that on the Helsinki device, top to bottom magnet polarity is reversed between the cylinder and the outer rollers and they roll opposite directions in an attracting magnetic field. Any electrons accelerated towards moving parts are seen to aid rotation through induction. The other way to reverse the fields in each shell would be to change the energy state of precession like with ESR. Alternate the energy state such that the electron shells precessing magnetic field actually flips over reversing its magnetic field. It is unknown what effect this will have on spin direction of a metal shell however.

Spinning Magnets

First a word of caution on spinning magnetic fields.

It is interesting to note that electric motors are now available that spin up to 200,000 RPM and no gravity effects have been encountered. Also magnetic flux fields still work, only mechanical limits are encountered. Transformers also take these moving magnetic fields to a much higher frequency and well into Gigahertz frequencies inside your desktop computer. Also atoms create the magnetic field in very small spaces by their electron orbits leading to the "intuition" that they are not large channels at their source. Flux does not seem to break down or split apart based on frequency, torsion, or size constraint. It is possible the "electric" or "time" fields of AG devices may be the source that destroys nearby electronics as well as air force jets that have tried to fly through them.

It was Einstein who suggested that antigravity may be produced from a rotating magnetic field. It is unlikely that a simple spinning magnet will do this, due to the motors now being built. It is also unlikely that Einstein viewed an electron as a small magnet spinning around an atom as more recent information has indicated may be the case, due to its 2 circular motions namely a "spiral" around a "circle."

It is more likely that antigravity may be produced from a particular motion or vibration of the Aether which can be set up by spinning magnets or by wave interactions of EM fields compressing the electrons helix and altering its curve. Simply placing magnets in a ring and spinning them has repeatedly failed to measure up. Although this will produce torsion fields, which may be a necessary component, something more is needed.

The Tornado Effect

Bringing order to chaos

A tornado that we think of may really be the center of a spherical torus which creates the system. The center is where the effect

Kosol and Koeun Noun Ouch Spherical Generator

moves fastest. All surrounding energies moving randomly are sucked in and realigned to an established order of spin. From a perspective of "moving Aether" the following has been conceptualized. Spinning a ring of very strong magnets in the presence of an Aether [2] high or low pressure may cause a vortex to form in the Aether that may become self sustaining at some critical threshold. As the speed reaches critical threshold there will be a sudden increase in RPM possibly leading to self sustained rotation and an inflow of energy from background ZPE. At these speeds it is said that the Aether[x] becomes like a viscous "elastic" fluid. Side effects of this phenomenon may be cars nearby stall, and other electrical fields are dissipated and de energized as energy is drawn inwards and reorganized. This is the Hollywood UFO effect. Also theories on dark energy or cold energy have been postulated.

At Helsinki the loading generators continued to work fine, and the system was slowed down using them to load the device to drop back out of critical ZPE speed. Also the Kosol device motors continued to work. The magnets of the Hamel device were also used to gate the device back down. There may be a proximity effect in action as well allowing systems inside the field to remain unaffected. The elements necessary may be "spinning magnets" as well as "magnetic or electric vibrations of a scalar nature" to feed the tornados energy until ZPE is reached. The Helsinki experiment, the only one to record actual data observed, indicates somewhere around 51.7 [tesla feet per second] was necessary to produce a self sustaining ZPE inflow, where the motors were disengaged and the device continued to run. I don't know how much of the field was opposing [tensor] and how much was attracting. Figures are based only on the presence of all magnetic fields present in the device and its spin rate, which are the main factors present. Once ZPE was reached it produced 8Kw of energy output and was only 19 3/4" diameter across the main ring.

The Roschin and Godin Searl Duplication

A study was done in Helsinki Finland of this device and the theory of TGD was the result. Thus the Roschin and Godin Searl disc duplication has been perceived as the "Helsinki Device" from this association. Any references to the Helsinki Device is an error but the slang has stuck around for some and appears in this text, as this document was the sole source of device information at the time.

The Helsinki device uses sets of opposing magnets which move between one another like a gear with magnetic teeth. The large cylinder possesses a uniform top to bottom N/S magnetic polarity which does not change as the cylinder rotates, thus no inductive drag is present to slow its rotation. The small wheels roll across the cylinder with opposing magnets one entering compression as one is leaving compression and these do not place a drag on the cylinder spin either as they tend to balance one another.

This pattern of fields produces a scalar or [tensor] interaction of pulses along the surface of the cylinder where the electron shell is expected to form. These magnets [opposing] may also have the effect of compressing the helix of the atoms particles pushing them towards ZPE because of the increased torsion fields produced which are not canceled. With the **Marcus device** scalar torsional energy is held in a section of the ring wrapped with a bifilliar winding, and then shot towards the core using high intensity pulsed stubs. The increase of torsion with repect to surrounding earth levels, the force related to time flow, can be thought of as an **Aether pump** or displacement, increasing the energy or decreasing the distance from center of spin of all matter in the area, as torsion will equalize to the surrounding sum of all matter around it over time. All these devices require an element of time to reach critical threshold, so the view of "**Aether pumping**" may be a useful model. In this case Aether, an obscure force, is actually identified as Torsion.

The operation happening between the opposing magnets can also be viewed as [Scalar ESR]. Atoms magnetic fields are tipping 90 degrees in two directions, or opposite directions, absorbing photons non coherently. As they tip back to center they generate scalar waves, releasing photons coherently but 180 degrees out of phase. These scalar waves then intersect and form interference patterns causing implosion or explosion forces at different points along the device which may cause rotational acceleration. Thus we have a source of non-coherent absorption of photons ejected as a coherent streaming Scalar pulsed field. This model may help explain why the Searl Disc turns and how it is able to reach OU [over unity power].

The Marcus Device

The Marcus device is an interesting system and not much is actually being shared on it. Marcus went private and avoids the press these days. A consideration of its basic elements show possibly the necessary fields present. At the center of the MD [Marcus Device] is an iron core layered with a thick dielectric and then a copper surface around that. The core is charged to 50 kv positive voltage. 6 rings with bifilar windings are spinning around the core on tracks, such that each one comes very close across its surface and then extends away in a circular ring. Along the rings are an arrangement of stubs pointing inwards which move past the core having normal coils wound on them. The first descriptions have 4 bifilar windings on a ring and 4 stubs setting exactly between them. The magnetic interaction on the rings creates strong scalar or tensor interactions, at the stub points between the windings, and for a very short crossing time the stub is fired with a normal magnetic field when it is closest to the core.

Here we see all the same fields present with the exception that the central iron core is not a magnet, but this field is pulsed in a stub 90 degrees to the scalar interaction on the ring, so the iron core is receiving magnetic pulses regularly and probably holds a pretty constant magnetic force, or a vibrating one. It was said that around the core was seen a brilliant electron glow. The device is said to be under study at a major university in the UK. Successive emails to them resulted in absolutely no information. They will neither deny nor confirm its existence. It is said to lift 2 tons. Descriptions indicated a fading out of the device under certain conditions [density shift], as well as AG, and a force field effect strong enough to deflect a hammer. Last I heard Marcus was having problems with the core iron breaking down, so it is probably that the magnetic pulses are jerking the magnetic field so fast it is damaging the metal. Or it could be the scalar interaction causing a torsion shock at the core causing a loss of tensile strength.

Magnetic Gating Technology

It has been shown experimentally on aa groups, that if a sphere is entirely covered with a magnetic strip, North Pole facing out all the way around, when the last section is placed all the magnets seem to shut down, as though their flux flow has been cut off. No gravity effect was reported. Also David Hamel's work mentions Magnetic gating having a gravity interaction. However Hamels gating technology seems to be performed only in one plane, giving him control only in one direction. His first craft shot straight up into the air and was gone, as witnessed by his neighbors. Just as to how gating Aether [3] magnetism can block the flow of Aether [4] Gravity is not well clarified. Possibly there is a saturation point where Aether becomes too far stretched to propagate Gravity. The spinning magnets pulling ZPE may be depleting it to local exhaustion, or this may be an incorrect idea.

Spinning Metal Cylinders, Spheres, and Pyramids

The original idea that SEARL started out using for his generator is still true. If a heavy metal cylinder is spun up at high speed electrons are hurled towards the outer surface. With the SEARL disc the presence of a large negative electric potential first appeared on the outer surface. On any charged sphere or cylinder the strongest voltage potential will appear on its outer surface. This is where it is most likely to jump off with an arc to any nearby grounded surface. Even though eddy currents may be formed deep inside the volume of the metal, the static charge will always move to the outer surface. In the case of the Kosol spheres the outer surface at the equator which is turning faster. High potentials were observed in every device thus far that succeeded at levitation, with the exception of the pyramid generator, which burned off its tips as it altered the gravity field and was seen to slightly fade out of time sync. The Marcus device, what little we have on it, was stated to break down its metal core over time due to the extremely strong EMF pulses being sent into it.

Possible Helpful Awareness About Working With Such Devices

In combining the three forces, [magnetic, electric, and scalar] some precautions may be in order.

Magnets could be mounted on inner surfaces of any metal spheres and all outer surfaces could be as smooth as possible. Insulating materials can be painted covering the inner surfaces especially magnets for closer tolerances, if it is thought that high voltages may become present. Outer sphere magnets may be blown apart if arcing does occur it will seek the sharpest points present on the outer surfaces to jump off. If outer magnets must be used consider coating only the magnets on the outer sphere with a spray on insulator for high voltage work. The outer metal surface may need to be left bare, to radiate electrons and fields. The outer surface is the critical surface for presenting electric potential and radiating electric fields. The distances between opposing magnets will regulate the strength and reach of scalar fields.

Section 3 [The Kosol Device Platform]

The spherical torus: [The self contained CU]

What is a Kosol Spherical Device:

The spinning of 3 or more spheres of magnets in the general pattern of an icosahedron, around a core. Successive spheres connected to opposite shafts of two motors. Allowing varying spin directions and varying spin rates to be applied. Magnet count is increased on outer spheres using the 3, 4, 6 pattern for flux intensity moving outwards. As many as 13 spheres have been conceived using the pattern 3,4,6,9,10,12,15,16,18,21,22,24,27. Successive spheres develop a conical shaped magnetic pulsing field tapering towards the center developing its tensor field at the point of crossing. This meets one of Beardens described models, a magnetic field increasing as it moves outwards.

The Kosol design offers an excellent platform for experimentation with spinning magnet designs.

A time gradient shift may appear at a location away from the spheres, also a gravitational effect. These having the properties of a corona at which point the presence of a magnetic control device may be effective for motion control.

Motors must be considered that will spin the weight of the spheres as well as overcome the repelling magnetic forces encountered during the orbit as magnets cross on alternate spheres. The first designs locked up due to insufficient motor power, and very strong magnets all in platonic patterns that were the same on each sphere. Alternate magnet placement could also insure magnets crossing in a pattern that would not fight sphere rotation if one set is entering compression while another is exiting it or if the scalar patterns are used to balance these forces.

Field Generation with a Sphere

The first field [magnetic]:

The first field must be an overall magnetic field encompassing the device. As electrons enter it with velocity they will be deflected or turned at 90 degrees to the flux field as they move through it. This will tend to move electrons in a circle around the spherical shaped device. This is a commonly understood effect in electronics, as a wire moves past the end of a magnet electrons are first pushed one direction, then as it crosses center direction reverses. Electron direction is perpendicular to motion of the copper wire but creates a sine wave as it crosses. Quantum physics tells us that if the first field becomes strong enough it will also align the electrons angular momentum and its magnetic moment into the same spin angle setting the stage for torsion manipulation. This field is not well understood at the atomic layer, and few have a good concept of why electrons stay in orbit if EM is the only explanation.

The second field [electric]:

A charge must be built up at the core of positive voltage gradient strong enough to propel electrons towards it into the first field. This field will contain the electron shell from flying off at random angles. This must be an electrostatic field because it must reach out further then the magnetic field. There is also a theory stating that electrons may couple [like magnets] to a surface as they pass it, conveying their energy of motion into the surface and bending them around a curving surface. This is

Kosol and Koeun Noun Ouch Spherical Generator

known as induction. The copper ring around the SEARL device may be operating this way as to electron sheets forming around it. This may be a factor for acceleration of the spinning device. However if electron speed is too high to move a copper ring or sphere alone, then as this electron shell begins to precess, or form a magnetic field that spins around its center axis, the magnetic field should inductively couple at a lower frequency and spin up the device. Device spin may be coupled to the energy state of the electron shell in this fashion, and device control will become regulating the density and frequency of the electron shell.

The third field [scalar]:

This is the field described by Wilbert Smith as the "tensor" field. It is formed between opposing magnetic fields and radiates scalar effects perpendicular to the plane face or surface formed between the opposing fields. If you want to gain a feel for this field take two strong magnets and push them together in opposition. The two magnetic fields do not cancel as current teachings would indicate, they create a strong tension between them which is easily felt as pressure. This becomes the active space for generating scalar waves.

The opposing magnets crossing will excite atoms causing the electrons in the outer atoms shells to increase in energy, jump off and be pushed outwards away from the small opposing magnets. If scalar transmission is present electrons may form at the crossing points of beams, as well as in intercepting matter. As electric force will reach out further then magnetic force so some electrons outside these magnets will be moved to velocity which will start to be effected by the other two fields and this process will feed the electron shell. It is not presently understood how these scalar effects work nor is the math in place to predict how much force it takes to start this process. This field is also seen to shift passing atoms magnetic alignment by 90 degrees which may release torsion into the device from the atomic particles if magnetic resonance as NMR or ESR is in operation in these atoms. At best we may expect to see a coherent vibration forming in the electron shell tapping the Zero Point Energy of Source and forming a vibration of **platonic** pattern. Whether the magnets actually need to be in a platonic pattern to induce this effect is not clear at present.

When dealing with spherical models it becomes apparent there are 12 obvious ways to position and move the tensor magnets of the third field. [Using a magnet pattern around a sphere]

1- North poles facing outwards - clockwise spin
2- North poles facing outwards - counterclockwise spin
3- South poles facing outwards - clockwise spin
4- South poles facing outwards - counterclockwise spin
5- North poles facing east - clockwise spin
6- North poles facing east - counterclockwise spin
7- South poles facing east - clockwise spin
8- South poles facing east - counterclockwise spin
9- North poles facing north - clockwise spin
10- North poles facing north - counterclockwise spin
11- South poles facing north - clockwise spin
12- South poles facing north - counterclockwise spin

Many combinations can be realized also by placing opposing magnets together at the points causing field extensions and trifield combinations, setting the stage for scalar transmissions.

All of the 12 methods should produce "spin" or torsion fields, as well as electrical fields as they move, creating both electric forces as well as scalar time variants and induction drag in metal spheres. Only the first four methods were initially considered.

It has been shown from tests with spinning cylinders that magnet placement as to polarity of magnets is not a factor in induction drag to rotating metal spheres of copper or aluminum, however angle of compression tensors may be an important factor. The Searl device aligns tensor fields at 90 degrees to the large magnetic field. Iron or steel spheres will not have any drag from

Kosol and Koeun Noun Ouch Spherical Generator

induction as their atoms easily realign with magnetic fields offering little resistance.

North poles facing one another may produce the opposite effect as south poles facing???

One polarity may compress and one may expand, this is unknown. The north pole is commonly thought of as the outflowing side of a magnet and the south pole the inflow. If this is true then North poles facing may produce a higher density connection and south poles a lower density connection, making stronger gravity a possibility as well. In a Kosol platform with three spheres we would expect one of each of these may form between successive spheres.

If it was our goal to emulate an atom we may think of electrons as aligned south up orbiting around a north up field, protons as north up orbiting around a north up field, and neutrons as two magnets in opposition held closely also orbiting the center field generating scalar waves.

Magnet Patterns

Magnet patterns may be altered to suit the experimenter. Ideally a magnet pattern that would offer a smooth rotation may be desirable as far less energy would be required to spin the device up. Inverse patterns have been speculated where magnets cover the entire sphere and leave holes in the platonic shape. The idea is vibrational control of the electron shell that forms.

ESR in the microwave region may be increased by cutting the magnets to resonant widths or lengths that correspond to their internal magnetic field, or slightly higher in frequency which should tend to make them become stronger as they gain energy.

NMR wavelengths are far to large to initially resonate in a small device, yet as a magnetic field increases, as may be expected these too may actually begin to resonate as density threshold is approached or crossed.

Discrete magnets arranged in opposing patterns are seen to offer **extended length fields**, both along the face where opposite fields interact, and outwards away from the center of opposition on the outside of the magnets. This field extension will allow even small magnets to create fields that will extend several feet. Magnets pulsing them at 90 degrees will modulate these lines of opposing flux and generate scalar waves. As a magnet crosses an opposing magnet, its field is pushed away and extends further behind. It also sets up the scalar face between the two, creating a much longer reach along this face. A magnet that interacts with another magnet also effects all the magnets interacting with with the other magnet in this fashion. A cascade reaction is possible that will travel through all the magnets, extending or contracting fields along the way.

The Atom Model:

Kosol offers these 2 magnet patterns as what has been communicated to him:

electron is isoca dedeca hedron,
proton is deca hedron, and
nuetron is isocahedron.

"The above reality is also true for the atom just like the below one in this statement is also true for the atom."

Electron shell = spinning dedecahedron or spinning isoca dedacahedron [highest frequency]
Proton shell = spinning isoca hedron
Neutron = star tetrahedron [two tetrahedrons in overlapping symetry]

This second model follows much of what we do know about atoms. The neutron having two magnetic cancelling parts possibly generating scalar waves. The electron having the highest natural NMR frequency. The Proton also spinning inside the electron shell at a lower frequency and spinning an opposite direction. Following this model would put the scalar **opposing magnets on the inner sphere**. The two outer spheres would have magnets overall up and down yet possibly with a tilted magnetic axis, producing wobbling magnetic fields as in the Hamel cones.

Here is the Atom Pattern:

The neutron [inner] sphere could simply be the heaviest magnets arranged in the two opposite polarity tetrahedron patterns, so close together that they create scalar interactions. Middle sphere lighter magnets North up [isoca hedron pattern tilted]. Outer sphere smallest magnets North down [isoca dedacahedron pattern tilted]

Scalar Magnet Patterns on the Kosol Spheres

The Earth Pattern:

Using Tom Beardens theories for scalar wave generation several other magnet placements have been recently purposed.

The "First field" may be moved entirely through the middle sphere using patterns 9 - 12, either removing the need for a core magnet in the device, or completing its magnetic loop. A scalar magnet pattern is formed between all the spheres. With this mode device lock up is nearly eliminated allowing for smaller motors to start the device. The middle sphere is configured with block magnets North up or North down, all the same at each platonic point on the sphere. If used, the core magnet is configured opposite forming a full magnetic circuit.

The inner and outer spheres are configured with magnets in opposition pointing towards the middle sphere and opposing the inner core polarity. The upper and lower hemispheres flips polarity on the inner and outer spheres to generate a compression field and mimic the earth as well. These magnets set 90 degrees to the middle sphere and cross them exactly at their center crossing points. All the magnets intersect in the platonic shape as the spheres are rotated. Spheres 1 and 3 are aligned and locked together turning one direction. Sphere two and a core magnet [polarity reversed from middle sphere] are locked turning the other direction. In this configuration the opposing scalar field is the pulsing field and the attraction field is running up and down the middle sphere and core simulating the earths overall magnetic field and all the earths major vortex points.

The first field now runs the complete cycle from its core out over the middle sphere forming the magnetic torus field.

The opposing scalar pulsing field intersects it at each point on the platonic pattern and along the core magnet.

Sphere lock up is reduced because as the opposing magnets on inner and outer spheres approach a N/S up/dn magnet on the middle sphere, both poles are equidistant, generating only an upwards or downwards pressure. By alternating hemisphere polarity on top and bottom of the inner and outer spheres, the up down pressure will be equal for the middle sphere and free rotation may be approached.

The scalar waves should radiate tangent to the middle spheres surface and intersect near the outer sphere where interference patterns, implosive or explosive forces, or plasma may appear. They may also appear tangent the core magnet surface. Changing the alignment between inner and outer sphere will tilt the scalar waves further outwards or inwards on the outer spheres altering where they cross.

Kosol and Koeun Noun Ouch Spherical Generator

The scalar core pattern:

An alternate pattern is used in the Marcus Device which accumulates energy around a center iron core. John Bendini has a model "The Scalar Beamer" which shows the configuration. This configuration beams the scalar energy directly at the core. The magnets on the inner and outer sphere are changed to the attracting field, and the middle sphere is altered to hold tripole magnets. That is two magnets pushed together in opposition, the center pole matching the outer spheres inward turned pole. Sphere two can use block magnets in the tripole configuration, and will be hard to keep in place while the epoxy is drying, opposing poles pushed together. If opposing magnets are positioned vertically, like poles outwards, turn resistance is kept low because as the spheres rotate, while there is repulsion from the outer sphere, there is attraction from inner sphere. Magnets on opposite sides of the spheres will balance the force.

General scalar patterns:

The ultimate is yet to be discovered such that motion is as free as in the Searl device, spheres floating. The nature of spheres does not allow for a spoked rotation so all will depend on the patterns of flipping magnets to balance all rotational forces and maximize the off balancing opposition pulsing forces. Scalar radiations are not well understood as of yet so caution is in order, as scalar beams if prolonged can overload the nervous system and cause damage. Keeping the scalar beams intersecting near the spheres is wise, as well as understanding which way they will project.

I have devised at least 12 patterns that show promise for a Scalar generating Kosol 3SD.

The Core Magnet Electron Accelerator

[The addition of a High Voltage DC impulse device to create the electron shell]

It has been demonstrated that by shooting electrons at one pole of a magnet the electrons will spiral as they approach and actually begin to orbit the magnet. If the magnet is a spherical shape they will tend to form a spherical shell around the magnet filling the containment field. Experiment proceeded in this direction hoping the natural field distances could be located allowing for the design of properly sized spheres to interact with a complete electron shell for the magnets power being used.

It may be experimentally possible to determined how to control precession of the shell to effect device movements. The simplest version of this is to electrically isolate the two shafts entering the device such that they form an electron spark gap and use a flyback coil output high voltage between them. It must charge the core magnet + and the opposing shaft and spheres -. As the voltage arcs electrons will be accelerated towards the magnet [hopefully with enough energy] to form an electron envelope around the magnet. The arc will also provide a wide range of electrical spectrum to excite the ESR frequencies of the magnet pushing it towards its high energy state. If "precession" of a standing electron shell can be mastered then the rest should follow. This allows for the design of the Kosol spheres to work from the inside outwards and removes the need for motors. The power unit can also be used to control the speed and output of the device if it is shown that electron density and precession frequency are the determining factors.

Although worthy of mention here, my personal experiments have shown that more must be necessary to form the plasma sphere then simply shooting high velocity electrons around a magnetic core. There is something deeper at work causing the atoms electron shell to form and hold its position. It may be that within the interference pattern of scalar waves crossing paths lies the true source of the plasma shell.

Kosol Sphere Analysis

Basic interaction for aluminum spheres:

Stability:

Consider spinning a single layered aluminum sphere, it's magnets with north poles facing outwards, inside a solid stationary aluminum sphere with no magnets. The magnets crossing the aluminum should produce eddy currents that will create fields opposing, or in other words "virtual" like poles radiating from the aluminum. The two spheres should develop a smooth torsion of pushing pressure between them, from the entire diameter of the outer sphere to the magnets on the inner sphere as well as an inductive drag.

This may become [extremely stabilizing] at every point where a magnet is located at high RPMs. Pushing the inner sphere centered within the system. With the magnets located along the areas of an icosahedron this will be a perfectly centering force. The faster the RPM the stronger the centering forces. The force will be stronger at the equator where speed is highest. It will derive its power from the torque on the motors and convert energy to a centering stabilizing force.

Now consider adding a set of magnets to the outer aluminum sphere with north poles pointing inwards, opposing the magnets on the inner sphere.

A similar opposing field should be produced in the inner sphere pushing the outer magnets out along the entire diameter of the inner sphere harder at the equator then near the points of axis except at the points of magnets crossing where we now have a scalar or tensor interaction. At the crossing points the opposing fields will bend each other away and may stop the eddy currents allowing only the two magnets to interact with a pushing field. This field may be stronger or weaker depending on RPM and thus more pressure and or physical vibrations may appear.

Placing the magnets on each sphere, both with north poles outwards, should produce a different effect of altering this opposing field, eddy currents will still oppose magnets, but now may be diverted through opposite magnets at the point of crossing creating only the expected reversed pulse of attractive pressure along those points. The pulse will be one opposite the repelling force along the rest of rotation so will produce a strong pull or vibration paralleling RPM x 10.

Thus we see the difference with reversed magnet polarity [tensor field] on successive aluminum spheres as being not vibratory rate, but in total reversal or increase, of the centering stabilizing force. As well as a deeper physical vibrating force with like alignments, and more overall opposing force for opposite polarities [more stable].

Field Considerations

Magnetic:

As the magnets spin around the axis poles in the pattern of an icosahedron, the reverse induced fields from eddy current activity, magnetic fields do not seem to add up to any unifying one field vectoraly that would reach out with any distance to encompass the device. Magnetic fields fall off quite rapidly as an inverse distance cube function. However on closer examination it becomes evident that as the magnets rotate they are also being vibrated in and out up to 10 times per rotation through a tensor area as they cross. As electrons may be pushed at a greater distance then the magnetic field around these magnets some electrons will be pushed outside their influence and begin to move within the other two fields if present.

Electric:

Eddy currents, or induced electric fields seem to be an active component for electrical activity inside the system initially. Although seen to offer [stability] in the device as inwards magnetic pressure between spheres, as electric fields propagate much further then magnetic ones this would indicate that induced field energy if present may move electrons towards the center of the system where they all cross. Also since the electricity is moving up and down perpendicular to magnet movements small longitudinal waves may be present moving along the outer surface of the spheres. The waves will be in phase since all magnets are aligned the same around the spheres with like poles turned inwards. As sphere groups, even or odd, are moving opposite directions the resultant electric paths will be a rough DNA, helix, or caduceus coil pattern as viewed from the side. This is the equivalent to coils of copper wound around a sphere with a caduceus pattern. This may produce [scalar] electrical waves known to penetrate through anything much like gravity. The waves will be moved into center and compressed at the poles crossing at the axis of rotation where they will terminate. We must also consider that inside all the atoms of the device is setting the source of gravity already in a state of operation, which does involve movement of electrons, protons, and neutrons as well at the atomic level.

Torsion Field [spin]:

The spinning matter will initially create two oppositely spinning torsion fields moving along the center of rotation upwards and downwards. All the atoms of the device will begin to time shift taking on a lower momentum and faster time frame as will become present in the Aether with respect to normal surrounding space. This may effect time by a small amount in the spinning components. 550 RPM / c, which is a very small amount, but an effect just the same. This torsion field will be confined to the matter of the device if it is not moved vertically. Over a period of time all the atoms in the spinning metal will begin to align to the new torsional energy state.

Torsion Electric interaction:

Here is where things get interesting.

The spin will also be throwing free electrons outwards along all rotating spheres as with the SEARL generator principle. Consider a magnet rotating at the core with both north and south poles in rotation. Although it's magnetic rotation may be producing strong voltages 90 degrees to its spin motion, sending some into the shaft, due to the fact that the shaft has thickness. These should cancel out as longitudinal AC eddy currents.

However the cores Electrons are being sucked outwards from spin away from center and may initially collect on its surface. Also, at the same time electrons in the outer spheres are being sucked outwards with a greater velocity because speed is higher with increased diameter. Electrons will slowly work their way from the central core through the shaft path outwards to the outer spheres and all the way to the outside of them, being more located along the equator regions of the spheres. Due to this interaction a "positive" voltage gradient should become present on the core. Considering the construction may isolate alternate spheres between the two motors, if not both grounded to one another, a second positive voltage may form on the next sphere out from core also appearing on its motor shaft. If motors are connected to the frame electrically it may now also start to become "positively" charged to a very high potential as electrons feed the spheres. If motors and shafts are well connected to each other, and kept insulated from frame ground, then a smooth voltage gradient will begin to manifest from core to outer spheres isolated from the frame. This effect if made to become strong enough through engineering will produce a total outer electron shell to form around each sphere, which may even begin to ionize the air. This electron shell appearing along the surface of a spinning sphere may now be present to interact with magnets passing by it and should begin to emulate the movements found in the electron shell of an atom. The electrons possessing a repulsion for one another will spread out over the entire sphere and be exactly where we want them to form an electron shell.

It has been suggested that a static charge setting on spheres in motion will appear as a magnetic field to any sphere moving relative to it. However if magnetism is viewed as a bidirectional torsion field this may not be true, and an overall magnetic field may not form from the static charge spinning alone on a moving sphere. It may be necessary to move the electrons through copper or aluminum atoms to produce a magnetic field, thus [current] may be necessary to produce a magnetic field, or the use of a permanent magnet to produce the first field.

The torsion electric interaction can be increased with the addition of a HV pulsed DC supply, and may even replace the motors if it is shown that the electron shells are capable of starting the device to spin from a stationary position.

Scalar wave interaction:

If the scalar magnetic patterns have been used then, as was predicted by Bearden, where the scalar beams cross one another can be expected manifestations of electron plasma, explosive or implosive forces, and possibly even psychic interchanges of energy, or time and gravity alterations. Magnet placement will be the key for discovery.

Spherical Torus Field [Magic]

Startup state:

The overall fields presented in the Kosol spherical torus device will be, magnetic, electric, torsion or time, and gravity. Eddy currents may be providing "stability" and they may be radiating both longitudinal and scalar or tensor waves during startup phase. Their strongest radiation point would be as they cross and scalar energy would be released into the larger spheres magnetic field.

The operation happening inside the magnets spinning around the spheres may be simulating electron movements around an atom and inducing such qualities of movement into all surrounding matter through the already present Gravity field. By having the time scale altered through spin, throughout the device "relative precession" of all matter effected will change in rate with respect to all matter outside the field. This would make gravity alteration possible at a local matter level only to a very small extent during startup on the spinning matter.

ZPE connection at 10 inches:

More importantly however will be the consideration of how things will react if a large electron layer is present on or around the spheres coupled with the positive charge at the core, at some point after the initial "Torsion - Electric" interaction.

Let's consider now we have a very strong layer of electrons on the fastest moving layers of all the spheres, their equator regions. The electrons are spinning with the spheres around in a 10 inch circle on the outer sphere, loosely connected to the atoms they orbit. The result is the same as with the atom, a magnetic field will be formed in the shape of a spherical torus around the device at 90 deg angles to the electron path. This field will now effect all electrons in the system tending to keep their movements lateral. It will not be particularly strong at 550 RPM but it will be present. Now as static charges continue to build inside this magnetic envelope, a very few excited electron begin to jump out of their orbits around the atoms they were spinning around.

Because of the presence of the initial weak magnetic field and the core magnet they will tend to move laterally. As they encounter the metal of the spheres they will induce reverse magnetic fields and tend to pull the spheres along with them, increasing RPM. As more speed is present now more electrons begin to be pumped to the outer surfaces and speed continues to accelerate. The typical surge of energy present in such devices now has manifest. If we are able to harness the near light speed electrons motion to move around our spheres then a stable atom begins to form. The typical blue corona will appear around the sphere, very faint at first. We have created a magnetic atom like iron or cobalt, with a diameter of 10", because most electron orbits are situated laterally around our device, they are all in the same axis of rotation. It now has an abundant electron shell moving at high speeds which we are vibrating with our magnets into the shape of an icosahedron. The whole electron field may begin to vibrate and pulse as magnets cross. The system enters the ZPE! A very large gravity spherical torus now forms around the device several magnitudes of force higher then at conventional electric speeds, with one layer present per sphere entering ZPE. After all spheres reach ZPE the final configuration of our device has reached full power potential. This would be expected to happen first at the outer shell and working its way inwards till all come up. The positive charge at the center of the device will pull electrons towards the spheres, and the outer magnetic field will push them inwards, being exactly out of phase with the inner magnetic flow by reverse induction. Also electrons in free orbit just inside the surface of any one metal sphere will encounter the reverse induction effects of stabilization that the magnets experienced during start up phase because they are in a spiral path forming small magnetic torus shaped packages as they rotate just as with the atom. The electron should be contained within the field and continue to tap ZPE as long as we choose to leave the device active, continuing to built until we stop or control electrons entering device orbits.

The "electric" field presented along the electron sheets formed just outside of our spheres now has the ability to reach out with an inverse distance squared function as with gravity. It will emulate the magnetic pulses presented with almost no resistance. Now the longitudinal waves, previously very weak in our magnets passing by them, will become standing waves top to bottom in the whole electron shell, which were present at startup only as eddy currents, but they will now be controlling electrons moving at near light speeds. The magnetic field will reach out a much smaller distance, creating the layers of our local systems corona and react with the electrons in rotation having a tendency to keep them harnessed around the device. The torsion fields present in the magnets are already receiving both circular spin, and spiral spin, induced from rotation around axis as well as magnet pulse. These fields will vibrate the whole area of the electron shell radiating out as gravity, just as in the atom. Since the electron shell in the atom is the point where all forces intersect, they can all be manipulated from this point now with relatively weak magnets.

Electrons are attracted to the core by a distance squared function.

Electrons act like magnets at a closer proximity, with an inverse cube function.

Electrons in a spiral path will act like magnets because they are moving in a coil motion.

Reverse induction is present and works for large magnets passing over a metal conductive surface.

In the Searl system electron shells must have been operating on the outer surface of the cylinder.

Kosol and Koeun Noun Ouch Spherical Generator

The electron shell will begin to precess or oscillate faster as its energy is increased and the device will accelerate.

Scalar beams crossing may create pockets of explosive or implosive force.

Implosive force dropping temperatures, explosive force causing heat.

Control:

The creation of a 10" atom now gives us the opportunity to study the physics of atoms, but more importantly when the magnetic torus forms it will enter ZPE as with all atoms. It may now be able to draw from the "magic" Aether containing ZPE directly from the center of spin of the electrons. We also may be able to determine which forces of "precession" control gravity and time where they come together at the curve of the electron shell at near light speeds. We would have a field with an area large enough to enter, altering gravity and time flow relative to the outside frames of reference. The fact that we have created a magnetic atom, lends to controlling it along is equatorial faces using magnets polarized, poles up or down, and along it's top and bottom regions with polarity. Now considering the interactions the pulsating magnets had initially, their effects at ZPE will affect the whole spherical field. They will still be relatively short distances, but now consider their ability to effect the whole electron shell. Since this is the layer purported to at least be partly responsible for gravity, we should be able to affect it for the whole system. It is through the electron shell that we can now radiate long distance effects, and through the small magnets we can alter it as if we were electrons. Now if we engineer alternate spheres to produce opposite polarized fields precessing opposite directions, we may become aware of two things. Magnetic fields are not canceled they merely bend and operate around one another. We will create more spherical torus fields at successive layers around the device. By having at least one layer in opposition this will produce the effect needed for external magnetic control! One field situated outside the other in a corona. At the balance point between them can be placed a magnetic device that should be locked into place. Moving one way will push; the other way will pull on our atom. This is the exact description of Kosol's reports from the very beginning!

Control may also be possible with external RF or even Scalar transmissions capable of effecting magnet ESR or NMR frequencies.

Analysis of parameters:

Altering the relative parameters of gravity and time:

It would appear that as torsion fields radiate only off the ends of a spinning wheel when it is given a movement 90 deg to spin, that on our atom altering or squeezing the spin path of the electron shell vertically may change our torsion gravity field. As we became weightless we also begin to move at a faster rate of time, decreasing our momentum as well relative to the rest of the universe. Allowing us to move around the universe with relative speeds very quickly with respect to those we left behind. We would be zipping around as the universe appeared to stand still. The opposite we encountered in Einstein's relativity where he had the moving ships time scale slowing down and its matter staying within this density.

[If moving towards the center of spin actually increases our time rate of flow as stated by Wilbert Smith]

The Searl device used opposing magnets at horizontal positioning to alter gravity. Pressing inwards they are reported to lower gravity, as they would lower its vectoral modulating component. This should also speed up time flow.

As the electrons or photons coming off atoms should already have the current position and time vectors for local space they should already be in sync. Our goal then becomes to engineer

Kosol and Koeun Noun Ouch Spherical Generator

our magnet positioning to accomplish the necessary squeezing to effect each parameter as desired. The devices thus far studied give us the patterns for gravity.

Opposing magnets placed pointing at one another both sides of electron shell should squeeze the helix along the equator of our sphere.

Power generation:

It may also be possible to create power generation using the reverse inductive qualities of light speed electrons from ZPE to spin a large metal wheel, without modifying its parameters of gravity or time. As electrons are relatively harmless little creatures and do not possess qualities of radiation that I am aware of other then light they should be easily contained within a spherical system of protection or the resultant magnetic fields. A set of cylinders could set up a ZPE field between them and the outer one should be able to spin up quite high in velocity.

This was the original intention of SEARL in the beginning.

Static electric orbital flow regulation:

As was stated earlier regulation of intensity may be effective through the control of the sheer quantity of electrons jumping to device orbits, and leaving atomic orbits. The release of energy will be electrons shooting outwards until the Electrostatic field is strong enough to contain them. At the point of stable containment a slow electron release may increase system output, a slow removal may decrease it. Thus regulation could be effected. If an HV supply is used to power the system rather then motors then it can be reversed to slow the system back to a halt and regulated to form a governor.

Intensity and Effects

Sphere thickness:

If the metal medium is made thicker more volume of magnetic field will interact. The sphere will become more stable in spin. A better conductor will produce a stronger eddy current, however there may be a trade off with "electric" field voltage if thickness approaches magnet fall off distance. If too thick some of the thickness will not be producing much current but it will be present to short it out, resulting in lower voltage for longitudinal effects. Thickness may be a balance of both effects, stability versus electric waves.

Sphere charging:

The addition of a high voltage gradient to the sphere surface, would require insulating it at the motor coupling flanges. If this was done it may radiate the electric fields much further through space and towards center. It is unknown if the spinning spheres would produce this voltage gradient of their own accord merely by insulating. However electrons do get thrown outwards from spin so this may in fact happen. Since there will be stronger waves near the equator insulating may not be needed. It is not readily evident what the actual effect would be. Maximum effects could be tested however with an external HV subsystem capable of 40 Kv used on insulated spheres. The supply must produce DC pulses, not AC and a method of arcing used, rich in harmonic output.

Spacing:

As the magnetic field is an inverse distance cubed effect, close proximity of spheres and magnets will be a cubed increase of magnetic force effect for a smaller distance, however less electron cloud space will result. 1/2 the distance between

Kosol and Koeun Noun Ouch Spherical Generator

spheres, will increase tensor field strength by a factor of 8. The closer they come the stronger the stabilizing field intensity will be and the stronger should be the electrical interactions, but the smaller the electron shell area will be traversed. The metal spheres must be arranged such that they do not deplete the natural orbital distances from the core magnet. They must operate around the electron shells but not deplete or short them out entirely.

The smallest distance allowable will be a function of just how high the voltages appearing on the surface of the spheres becomes. Any leakage from arcing between the spheres may lower the torus electric field which is a function of longitudinal currents flowing in the spheres. Teflon insulating coating or other high insulating material on the inner sphere lining, especially over the magnets may improve distance tolerances if such arcing should appear. Unfortunately if it does appear, magnet surfaces will get hit first. A Teflon layer was used in the Searl device between the inner cylinder and the outer ones. It was referred to as the regulating medium limiting electron flow outwards but may have actually functioned as a capacitor.

Magnet polarity and pattern:

It was discovered early on that placing powerful magnets in a platonic shape causes a locking up of spheres and takes more powerful motors to overcome this resistance to motion. Inverse patterns have now been suggested that should reduce this greatly. A pattern may yet to be discovered such that will form a platomic vibration with magnets that alternate crossing points to reduce sphere torsion power.

Construction

As to construction materials:

The use of fiberglass as a sphere material has been suggested as well as copper, aluminum, iron, steel, and brass, [or even possibly a laminate of aluminum foil layers separated by a ceramic past mixed with miniature spherical plastic spheres that are 2 millimeter in diameter].

Fiberglass:

Should result in very uniform Magnetic fields, a dynamic electric field will develop only along the distance of the metal shafts turning the device. Because the magnets spin around the shaft in opposite directions, this voltage produced along the shaft could vary, canceling out as longitudinal eddy currents. Static charges however may align along the outer surfaces from electrons thrown outwards, more research is needed to determine static charge effects and electron layering distances from the core magnet. These static charges may end up on the motors or frame so need to be considered. If they are necessary for the torus to form, they may benefit from some help in forming more quickly. The Van De Graph generator principle may help to speed this process. The rotation is already present, the addition of few strategically placed materials could increase this effect.

Electric forces will now have no medium to travel through or become shorted out in the static field on the outer surface of the sphere, so resistance to tilt effect may be lower.
"Eddy current stability" will be less, as fiberglass in an insulator, present only at the crossing points of magnets. This will reduce motor loading but sphere balancing will be more critical. Balancing these spheres may become difficult at high speeds until ZPE is reached, where the presence of smoothly flowing energy becomes stabilizing.

Kosol and Koeun Noun Ouch Spherical Generator

Copper:

The development of eddy currents is present whenever a strong magnet is moved along a conductive metal surface. For the choices of metals presented copper is only surpassed by silver in its conductive abilities and is twice as conductive as aluminum. These currents will flow in the direction outwards, or inwards, perpendicular to the motion of the magnet as it passed by. If all the magnets fall into basic ring patterns, then rings of eddy currents should be expected to develop along the latitudes of the sphere. They should move both outward then inward as a magnet passes by generating an EM field which will oppose the magnet that induced them, this is a [stabilizing effect] that increases with metal thickness. They should appear as strong AC currents and magnetic fields that move along with the magnet that produced them. I calculate most of the simple eddy currents should be between 45.8 Hz to 183.33 Hz at 550 RPM assuming a dodecahedron magnet pattern on all the spheres with only one motor on. With both motors on it may double to 91.6 Hz to 366.66 Hz. More magnets and more spheres will result in a higher frequency for a given RPM. Eddy currents will also release heat, as well as stabilizing magnetic fields. As to human danger eddy currents may only cause a lethal problems if the sphere accelerates beyond, 800,000 RPM where dangerous microwaves would be the main radiation. Also static potentials may appear with a possibility of lightening strikes between any metal surfaces on the device and ground, or between the framework and the spheres. A complex merging of harmonic fields in the 45 to 366 Hz range could produce higher frequencies, but probably nothing to worry about.

Electron cloud interactions could be expected to be at maximum operation for copper spheres and if electron or magnetic fields are used solely instead of motors this may be the best choice for a sphere material.

*** The Searl device had a copper ring and magnet rotation appears to rely on the magnet mirroring effect of copper induction to some extent.

Aluminum:

In the Russian research of torsion fields, it was discovered that aluminum was unique in that it had the ability to reflect some radiated torsion fields, that would easily travel through copper and other substances. It will also conduct eddy currents quite well and used to be used as wiring in homes. It's resistance to electricity is higher then copper, so eddy currents will be reduced a little, and thus EM fields may be increased with a lower stability factor. When eddy currents are present the metal can be repelled along with the eddy currents produced in it by an AC field. The effect is amplified when the metal is cooled with liquid nitrogen. Spinning aluminum by one pole of a magnet produces a repulsive force.

Iron, Steel, Nickel:

The use of iron, steel, or nickel in sphere construction will introduce a new component that of magnetic flux transfer between the magnets located along its surface as well as extending the core magnetic force outwards. Much as a transformer core does, it will spread out the magnetic field of each magnet. If all the magnets along the surface of one sphere are polarized the same direction, it is possible that a total magnetic gate could be created, that would basically shut down all the flow of flux into or out of the device. This may have the effect of turning off all the magnets. However if magnets are kept far enough apart, or are reversed in polarity along the surface, then the only effect will be that of focusing the magnetic fields onto the spheres surface, and creating more wider magnet area.

Eddy currents will be altered and drag from induction will all but disappear with iron, as the atoms of iron freely rotate to follow any external magnetic field and do not produce a perceptible drag from magnets in motion around them.

A sphere of these material may be considered if it is desired to extend the core magnets flux outwards much like in the Searl device where a magnetic ring is used. In a scalar magnet arrangement this material could be used in the middle sphere forming a complete magnetic loop with the core magnet.

These elements also allow for the sphere to be magnetized and become the source of the first field as in the Searl Device. A magnetic ferrite coating could be added inside a copper sphere as well and magnetized N - S up - down, to provide the first field rather then using a magnetic core.

Brass Bronze:

Should fall somewhere between iron, copper and nickel, which is basically what it is made of. Although magnetic qualities are greatly reduced eddy currents will still flow. Gained weight would seem to make these the least desirable materials yet less likely to rust or corrode.

Motors:

Kosol has indicated motors that spin up to 5000 RPM may be preferable. DC motors that may also be used as generators would be preferable as they could also extract energy from the ZPE. Shaft size was recommended at 3/4 inch or more, this is very large and should be able to handle the stress of large torsion. It was discovered that poorly designed motors can be effected by the magnetic flux entering a motor through the metal shafts. The magnets on a sphere spinning can induce a magnetic field in the shafts that will move all the way into the motor windings and saturate them, slowing the motors. Extra care should be considered when procuring motors, ones that have internal magnetic isolation.

Notes and Suggestions to Expand Awareness

Strategic placement of insulators:

By using full multiple spheres the current necessary to form the torus may be readily available at lower voltages. However, it may be desirable to insulate the metal spheres from the motor shaft in order to allow the build up of an initial potential to form the torus more quickly. Especially near the equators where magnets are moving the fastest. Also coating the inner sphere and magnets with a high dielectric insulator may prevent arcing between spheres, which may drain the magnetic field or the torus energy.

The SEARL disc used a metal cylinder as did Helsinki. At Helsinki the cylinder was insulated from ground so an external voltage gradient could be applied. It was not necessary to achieve critical ZPE, however when used it altered the gravity differential. The Searl disc was said to produce lightening strikes, it is not mentioned whether this could be grounded out and not affect functionality. The MD RP uses a copper sphere insulated from its iron core at the center by 1/4" high dielectric material.

All known devices produced a high potential on the outer surfaces of the inner element which moved outwards, as electrons were hurled outwards. This is an unknown and may only be a safety issue to prevent large currents from reaching the motors and jumping back into the power source. Insulators can be added to the legs at the floor to prevent unexpected arcing to ground, or grounding wires could be attached to prevent arcs from blowing holes in cement floors. Standing the unit on plastic egg crates may be an easy alternative if necessary. If a HV supply is considered with core arcing electrodes, then the addition of a Plexiglas shield can be placed between experimenter and experiment during trial phases.

Kosol and Koeun Noun Ouch Spherical Generator

Magnet placement:

Also as electric potentials always form on the outside of a metal sphere it is my calculation that magnets would be better utilized if placed on the inner surfaces of metal spheres, and all outer surfaces be kept as smooth and perfectly round as possible. Far less arcing should ensue with this construction. Also magnets should remain more stable as centrifugal forces will seat them rather then tear them off. And electric arcs may not blow them off as bad.

External container:

It may be advantageous to add an outer stationary sphere, to provide stability to the device at high speeds. Providing an external completely smooth interaction of compression which should reflect all the way to the inner spheres. Also this sphere can be insulated from the shaft and allowed to charge up creating an outer layer of electric flow for the torus to form or it could be grounded if far enough away to prevent arcing. This container may also allow for a better anchoring if used to mount the device inside a craft.

Magnetic braking system or speed governor:

The conceptualization of a runaway device presents a safety challenge.

An outer shell may be conceived of a flower shaped design or simply a few petals of a flower which could be raised or lowered, moved in or out, to place drag on the system. Using lateral magnet polarities which will resist the rotation. A very solid row of strong magnets with correct polarity would be mounted top to bottom along the curve of the partial shells with poles pointing east and west. This should be effective for copper, aluminum, brass, or steel spheres.

A shorting bar, switch, or high resistance resistor may be positioned to drain off the electron layers forming around the spheres and short them to the core. The only way to harness a runaway scalar device may be skewed magnetic fields to alter the scalar radiation angle, or scalar beams that effect resonance at the correct frequencies. A scalar beam made with similar strength magnets, or a smaller sphere radiating an opposite polarized field.

Disclaimer:

This is my present developing theory as to the design of antigravity devices. It is not to be considered as written in stone, much is based on observation of claimed working devices, which have not been widely proven to exist, witnessed only by a very few. [Subject to change and modification as experiments will continue to improve the design.]

Section 4 A Search for the Perfect ZPE Device

Functionality:

The unit must induce near light speed atomic particle forces at some point after startup phase. Harnessing the increased magnitudes of the atom drawing from source.

It must do it in a safe manner, non destructive, and controllable.

It must not cause destructive effects to surrounding systems or humans.

Starting methods:

1- The application of external energy. Using physical motion such as a motor(s) or High Voltage DC pulsing or arcing devices The use of external magnetic fields or other sources to achieve self sustained motion.

2- The use of perpetual motion or ZPE from the slow buildup of an oscillating magnetic field like in the Hamel cones

Device layout for a self sustaining operation:

1- The use of a multi layered vortex system, utilizing the spherical torus form which offers total directional control and self contained effects as found in operation at the electron shells of the atom.

2- Devices using a single vortex operation layer, or bidirectional nature.

Control methods:

1- External magnetic gating of energy inflow for control and shutdown.
2- Adjusting the active element input carrier source, ie static electric orbital flow rate or electron cloud density.
3- Motor or resistance loading, tapping power off using a governor device or generator.
4- Scalar beams or skewed magnetic fields.

The introduction of magnetic bearings:

Brings four things into the system:
1- Frictionless movement, and higher RPMs possible, total floating magnetic Spheres.
2- 4x energy compression to the pole being compressed.
3- Magnetic gate at the points of shaft entry.
4- We are now faced with finding a new method of spinning the device, magnetic startup or magnetic spin.

Notes:

Interesting to note that on a rotating sphere the magnets at the equator are moving faster then the magnets on the Polar Regions. This is the opposite effect observed in a tornado. However the resulting effect set up in the Aether may be much different.

More on Magnets

[As to fully floating spheres]

Pyramids and magnets:

If you move a magnet around a circle or ellipse pattern laid at the center of the outer face of a pyramid such that it moves all the way to the sides [not top or bottom], at the center of the base of the pyramid you get a rotating field. As shown by placing a compass at that location. Moving the circle pattern up very near the top of the pyramid, will still produce a complete rotation of the compass even though the magnet is only moving a few inches. This seems to be because of the angle which a magnet emits flux patterns. My experiment was conducted with 2000 gauss magnets on an 11" hollow pyramid. If you spin at all four faces they should all interact at the base. By spinning a properly aligned and polarized ring of magnets along the faces of a pyramid, you should be able to spin a magnet on a shaft at the center N/S facing outwards. [Placing half the ring with North in and half with South in may produce a smooth turn.] This may provide a method of startup for inner spheres without using motors.

Octahedron:

Rings with magnets facing in spinning around the faces of an octahedron should spin a magnet at the center in any of 3 planes [through the 4 point centers]

Post Analysis

During the creation of this document it was realized by the compiler that when you get to the electron shell of the ATOM, the smallest unit we can see, using the background ZPE as its total power source, you have reached into all the mysteries of, electricity, magnetism, gravity, time, and [Source] energy. When solved you do not get just one solution, you get them all. It comes in one package, is the basic structure of this universe, and never needs charging. The relationships may be determined from the geometry existing at this level and the principles we have already known.

The key principles in this research, in a logical manner, one by one, have linked and shown the relationship between all these forces. Only the "comprehension" is necessary to put them into practice.

The solution to ZPE, and [time - gravity] relationship becomes "intuitively" understood. The message that is presented with the introduction of this secret, merely, Comprehension. Without it, the recorded data is worthless.

The data within this document did not come from the compiler of information, it has come from the dedicated individuals who have blindly fumbled and haphazardly succeeded at producing ZPE but never realized what they had done or how. Within is the understanding of the SEARL disc, the Kosol spherical device, the understanding of harnessing the sheer magnitudes of power, along with the beauty of the Atom. Its wondrous function as a stand alone unit of natures most unwavering source of total eternal existence, because within it is contained infinity. Within infinity, there are No Limits! It makes sense that a device intended to harness such universal atomic forces would be designed after the atom and take on the shape of the spherical torus of nature.

Electrical and Mechanical Analysis for Kosol Device

Note that this is not a definitive conclusion, much more will follow as building among the groups develop.

I have based this work on Lacosta's configuration with 5 multilayered spheres and the electrical part done by me and Vince.

I know that you probably have already computed inertia momentum for the spheres and total mass, but I have decided to separate the circuit powering the motors, so I will do this separation:

Circuit 1: Powering 1, 3, 5 multilayered spheres
Circuit 2: Powering 2, 4 multilayered spheres

The reason why i have separated the circuit will become apparent further in this document.

Each sphere is made of different materials:

Bi, Al, Bi, Dielectric, Bi, Al, Bi

We have 5 of these spheres with decreasing radius following platonic math. Also magnets follow the same configuration, a dielectric between 2 magnets.

Inertia momentum for each single sphere is $J = 2/5 \, (m R^2)$ since the sphere is made of multi layers we have to compute the total inertia momentum for the entire multilayered sphere:

R1 = radius for Bi shell
R2 = radius for Al shell
R3 = radius for Bi shell
R4 = radius for Dielectric shell
R5 = radius for Bi shell
R6 = radius for Al shell
R7 = radius for Bi shell
R8 = radius for distance from the rotation center for the entire first multilayered sphere
(Look at the drawing made by Lacosta for reference.)

Computation for each shell gives 7 quantities:

$J1 = 2/5 (m1 (r1-r2)^2)$
$J2 = 2/5 (m2 (r2-r3)^2)$
$J3 = 2/5 (m3 (r3-r4)^2)$
$J4 = 2/5 (m4 (r4-r5)^2)$
$J5 = 2/5 (m5 (r5-r6)^2)$
$J6 = 2/5 (m6(r6-r7)^2)$
$J7 = 2/5 (m7(r7-r8)^2)$

Total inertia momentum and mass for this sphere are:

Jt = J1+J2+J3+J4+J5*j6+J7
Mt= M1+M2+M3+M4+M5+M6+M7

This computation needs to be done for each of the 5 multilayered spheres.

So we will have Jt1, Jt2, Jt3, Jt4, Jt5, and Mt1, Mt2, Mt3, Mt4, Mt5

Total mass and Inertia momentum are:

JT=Jt1, Jt2, Jt3, Jt4, Jt5
MT=Mt1, Mt2, Mt3, Mt4, Mt5

Kosol and Koeun Noun Ouch Spherical Generator

Circuit for the Kosol device:

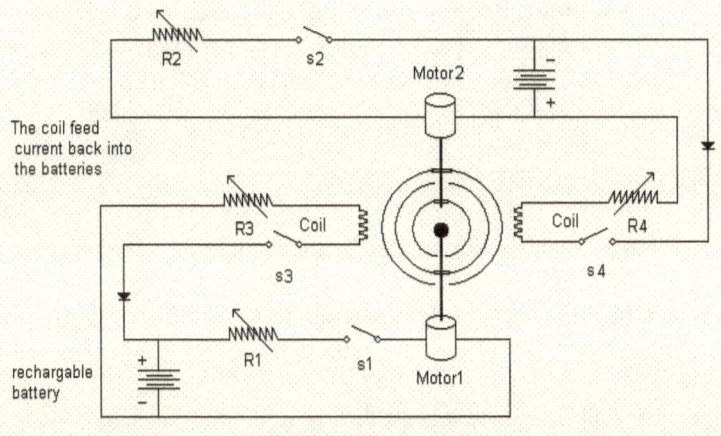

Note for Lacosta :

I have seen that total mass for your device is 233kg, with that mass you need industrial grade motors, you also have to choose according to your needs, if the motor will be AC or DC, since this device is outside current electrical assumptions, at the moment I don't know which result it might cause if you attach an AC motor to the power grid.

I will make the assumption that the motors are DC. In my device the total mass is less than 2kg, and I have two DC motors which unloaded have a regime speed of 21000 rpm.

Coils are put at the equator of the sphere since magnets have the maximum linear speed, the inductance for the coils is Leq and is equivalent with the series of single inductance coil:

Leq = L1+L2+.......

Building notes:

Every DC motor has 2 parameters K and b.

K is a scalar value indicating the electrical counter-tension during duty cycle for the same motor, b is a parameter depending on the attached mass and friction at the connection points. You should know both these values are given by the constructor.

Equations for the first motor are:

E is voltage, Om is angular speed, a is angular acceleration

E-iR-KOm=0 (electrical part)
Ki-bOm=Ja (mechanical part)

Since a = d Om / dt, we have a differential equation which needs to be solved:

E-iR-KOm=0
Ki-bOm=J (dOm/dt)

Solving these two equations gives:

d (Om / dt) + Om (k^2 +bR)/ RJ - k E / Rj =0

Let's use some constant for ease of computing:

C1= (K^2 +bR)/RJ
C2=K/RJ

The differential equation is:

d (Om /dt) +c1 Om -c2 E =0

To solve this equation we need to know differential math. I did all of the computations for you, and after computing the solution and relative constant at initial at 0 sec we have:

Om(t) = (KE/ (K^2 +bR)) (1 - exp (-k/Rj) t)

This equation is divided in two parts, a regime state and a transient. The transient is the main reason why I have split the circuit into two parts.

Transient state will give indication about how slow or fast the system will reach its working stable condition. Since two different motors are powering the circuit, and each motor has a different mass to rotate, we will have two different transient and regime states.

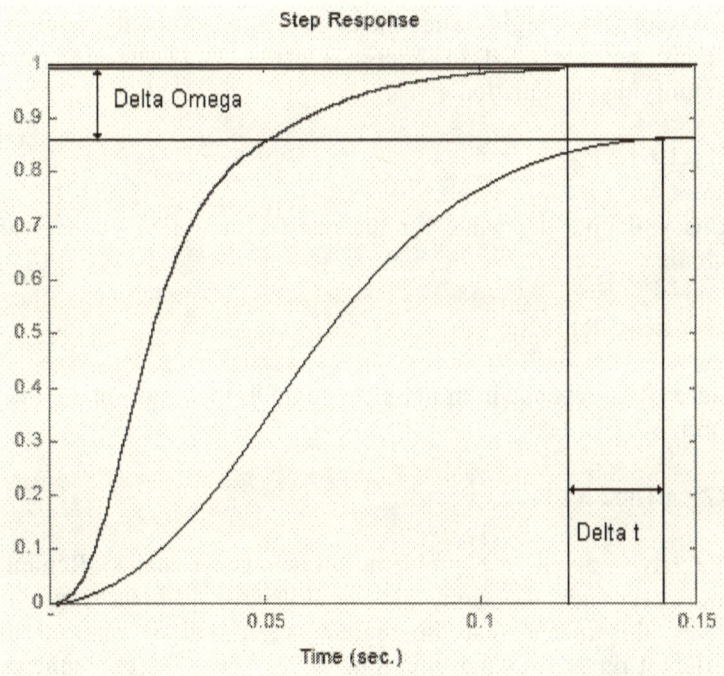

Top curve is transient for first motor and the bottom one, for the second. Both have different time constant and regime state. This means that one motor, due to different mass to rotate will reach different speed of rotation in a different time. We need to solve this problem.

First of all the two stable angular speeds are:

For the first motor:

Om1 = K E / (K^2 +b1 R1)

And for the second motor

Om2 = K E / (K^2 +b2 R2)

We want that at regime state the angular speed to be the same so simply put we need to impose Om1 = Om2. Solving and simplifying the equation gives:

b1 R1 = b2 R2

This will give the value for the resistor to use in the second circuit:

R2 = b1/b2 R1

Since b1 is dependent on mass we could use the ratio in another form:

R2 = m1/m2 R1

We have solved one problem but another remains. Regime speeds now are the same, but we still have different transient states, this means that the regime speed will be reached in different times, although this time is very little the two spheres might be phased out during rotation. Fortunately because of magnetic interaction between the spheres the phasing out will be more and more less as the rotation proceeds.

If you still want the solution here it is:

Time constant for the first motor is:

Kosol and Koeun Noun Ouch Spherical Generator

$t1 = R1J1 / k$

For the second:

$t2 = R2J2 / k$

$Dt = t1-t2$

$Dt = (R1\ J1 / k - R2\ J2 /k)\ 0.63\ sec$
(0.63 sec is a constant, and normally is referred as the time that is needed for a system to reach stable state)

So, if you want to nullify the difference in time between the two circuits you need to start the second one with a time delay of Dt.

Charge measuring:

Charge in this device builds up, we still don't know how much this charge will be. If the device follows the lifter tech or Brown effect, we at least must expect 17.35 kv to see noticeable antig effects. To measure you can do two things, build a circuit of your own (able to measure femtofarad to picofarad) or buy an high impedance capacitance meter.

Since you want to measure charge for the shells you need to put the anode and cathode to close the circuit on the outer and inner shell for each sphere. The device is rotating so you will need mercury contacts or brush contacts (subject to consumption). You need to place firmly the contacts between the spheres and make all the wiring connections pass through the inner shaft which needs to be hollow.

If you want the schematic for a capacitor meter here it is, but I advice you to use a professional one since it can be used with much higher precision and could measure much higher voltages.

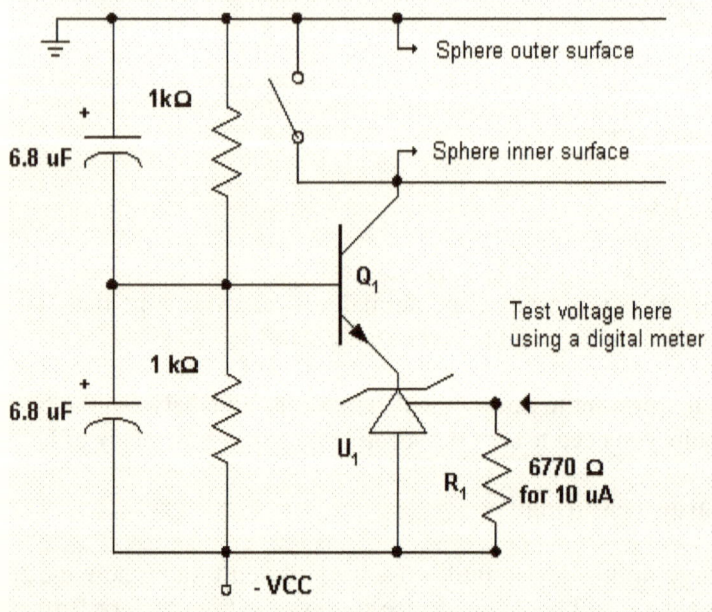

Vcc is at -5v
U1 is a Schoktty diode voltage triggered
Q1 is a npn bjt normally they have a very low parasitic capacitance between collector and base.

Working with the device:

Power the two main circuits, keep the feedback coils off and let the device reach 550 rpm with a stable rotation. Turn feedback coils on and after a certain amount of time the device should sustain itself and main powering circuits could then be turned off.

Closing Comments

The secret to my device is that the spherical device is a capacitor and is also a generator, meaning that it is constructed to mimic the capacitor component that you find in a capacitor circuitry with dielectric and aluminum plate (spherical plate.) And is also a generator, meaning that it has magnets that are put on. These magnets on the surface and inner surface (you can have magnet on both surfaces or just only on one surface is fine, no limit) of the spherical aluminum plate are used to create lens law by rotating the magnetic field to cut across the device feed back copper coil. This coil feed back circuit is used to feed power back to the motor and rechargeable battery. This is the generator part, it gives itself power from the lens law affect and you can siphon this power to run your house and car, etc. Now the dielectric which is barium tenate or just plastic alloy or some sort is to help isolate the spherical plate from the shaft of both motors and also isolate the spherical aluminum plates from each other.

Now as you can see the device components come together and you get a capacitors like structure. Now the voltage ionization charges come from the magnetic field that cut across the coil and the aluminum spherical plating at the same time. This device stimulates it own power source as it spins. The charges build up automatically like a capacitor because of the isolation of the spherical plate from each other and the shaft motors by the dielectric materials.

Now the device is reaching new charges and ionization as it spin and the device will glow automatically like a UFO. Like the lifter technology the spherical Kosol device is the same constructs except unlike the lifter the Kosol sphere device creates it own power source and supply as it spins. The Kosol spherical device increases it own ionization, and the device lifts up into the air and unlike the lifter that requires 50k voltage from an outside power supply to help it lift, the Kosol device creates it own power to help it lift. Because the lifter technology is also a capacitor it has aluminum foil and there is gap between plate a and plate b, except all of the lifter plates are in the shape of a triangle that is the different. (By the way Timothy Ventura lives here in Seattle where I live also.)

As you can see my spherical device is the same as the lifter except that I give it a spherical shape and I use the faraday mono generator and lens law concept by adding magnets to my spherical capacitor to generate electrical power from its own self rotation. The feed back copper coil circuit comes in handy in my device. See my technology is not that advances as one thinks. It is simple and very effective, that's all. Just like the lifter, all alien UFO or Vienna (ancient Hindu, Atlantis or asvin flying saucer) use the same concept.

Thanks in advance, now you have the secret and there is nothing new, I just connected the dots for everyone.

Also, as well, you can use the dielectric material to put it in between the magnet and the inner surface and outer surface of the aluminum spherical plate and it can be in a coin shape.

P.S. The time craft also uses the sphere technology. It has eight Tesla coils, four on each hemisphere, two motors, one for each hemisphere. The time craft uses the single hemisphere or multi hemisphere version as it drives. The hemisphere is constructed like the hemisphere in the Kosol and Koeun spherical generator device except that it uses bigger dimensions to fit the purposes of the time craft. Also, on the sphere device the builder can use a barium tenate coin (thin dielectric barium tenate coin that is) and put this dielectric barium tenate in-between the magnets and the sphere for every magnet throughout the inner surface and outer surface of every sphere and also on the Kosol and Koeun spherical generator device.

Illustrations

Kosol and Koeun Noun Ouch Spherical Generator

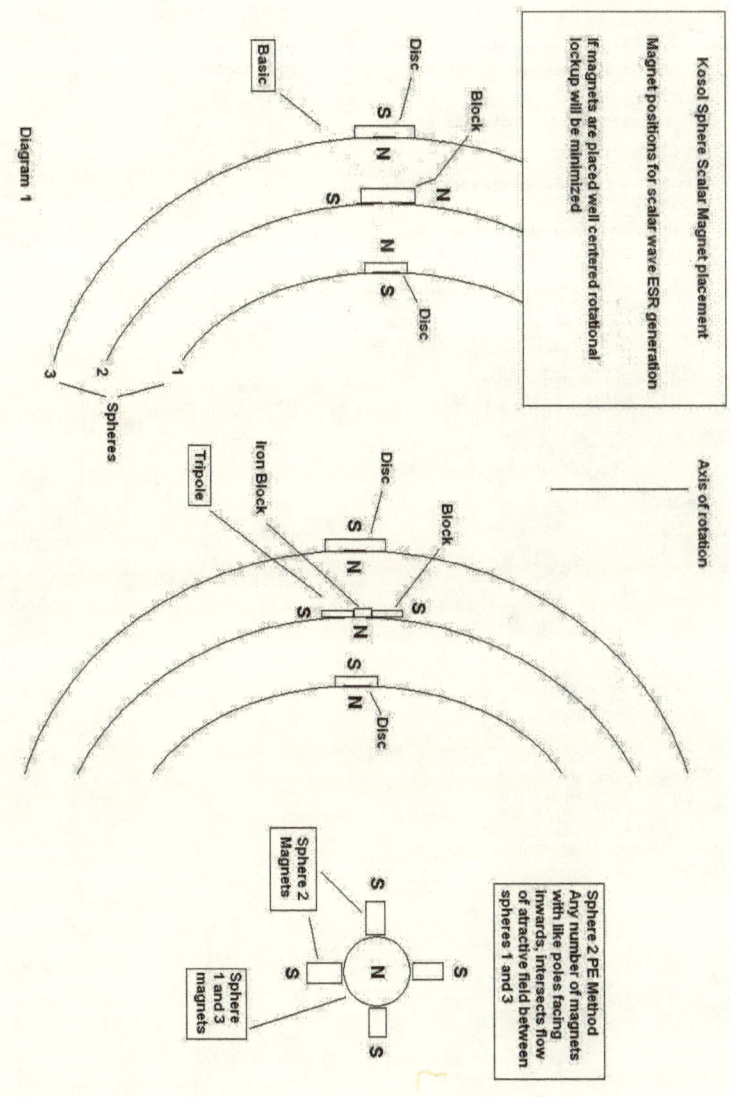

Diagram 1

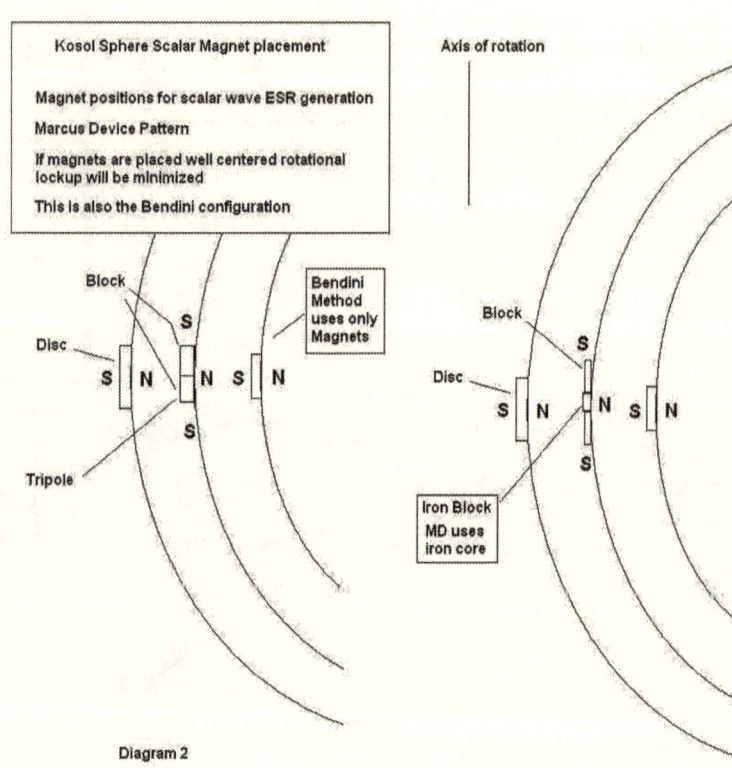

Diagram 2

Kosol and Koeun Noun Ouch Spherical Generator

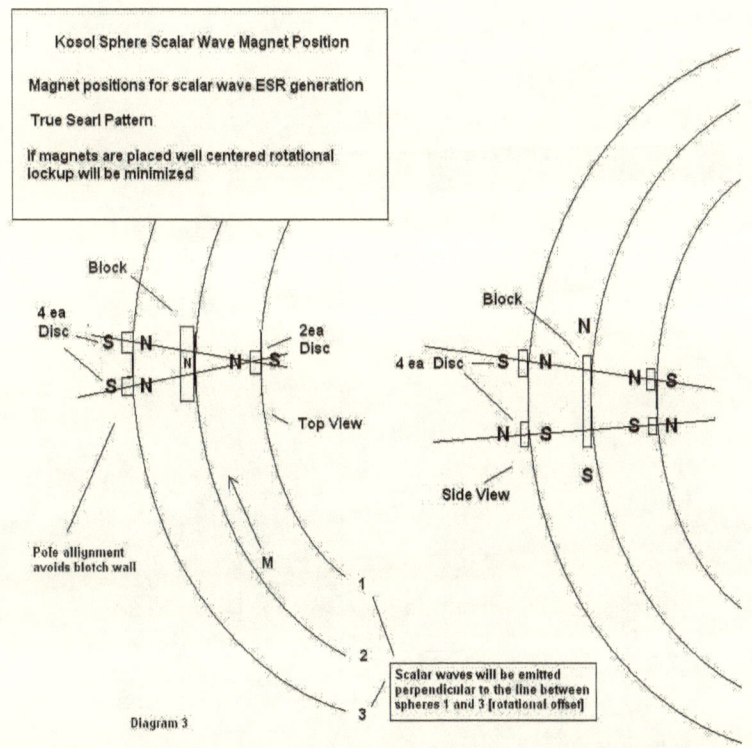

Diagram 3

Block Magnet

Shows areas where atoms are most easilly tilted

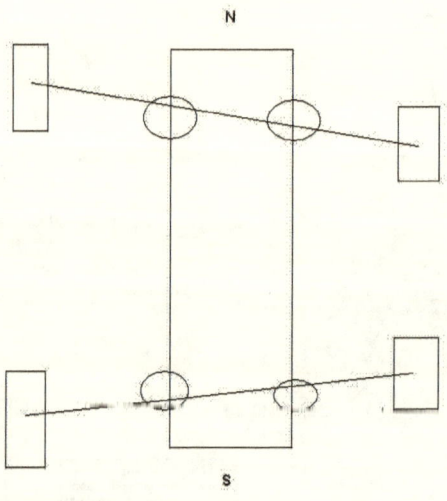

Diagram 4

Kosol and Koeun Noun Ouch Spherical Generator

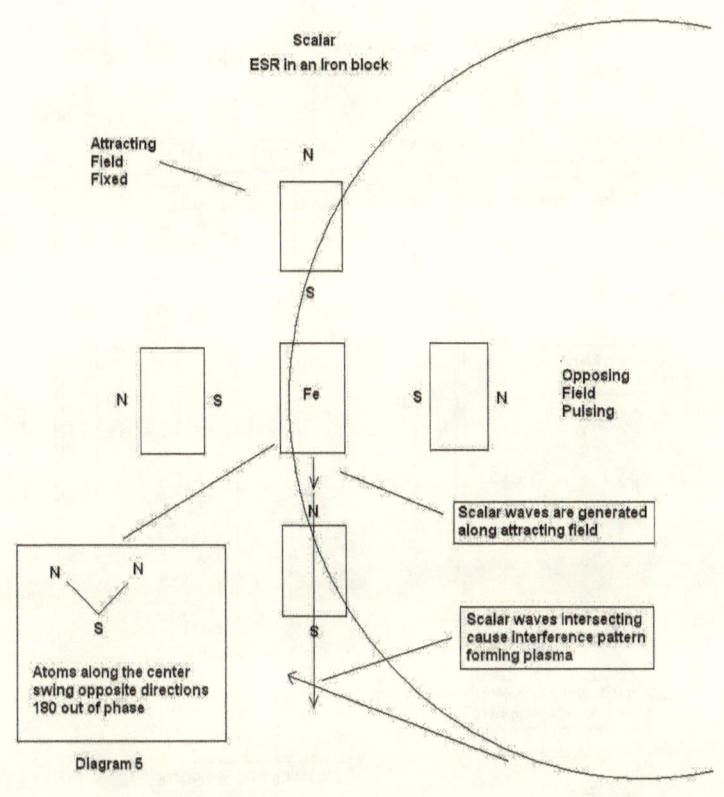

Diagram 5

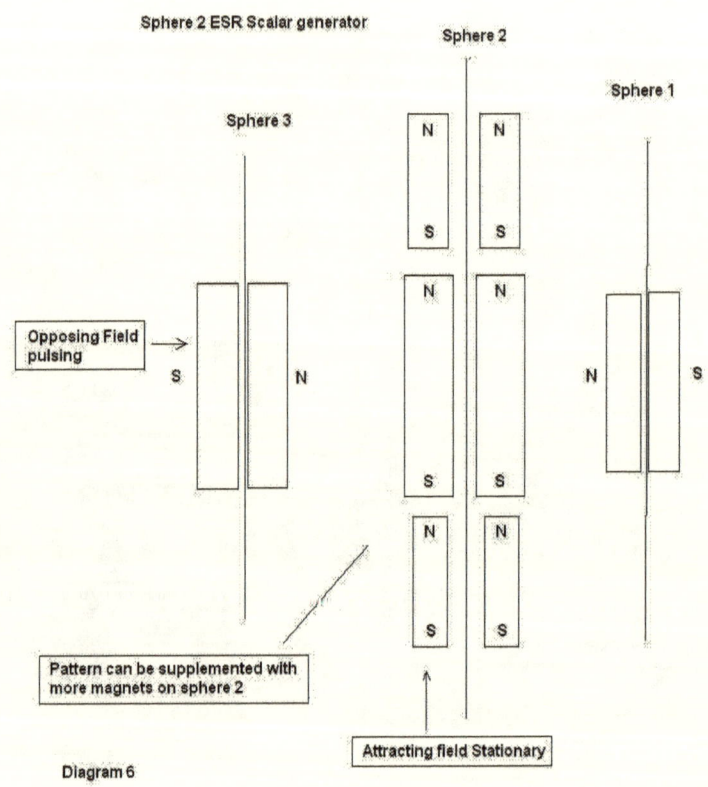

Diagram 6

Kosol and Koeun Noun Ouch Spherical Generator

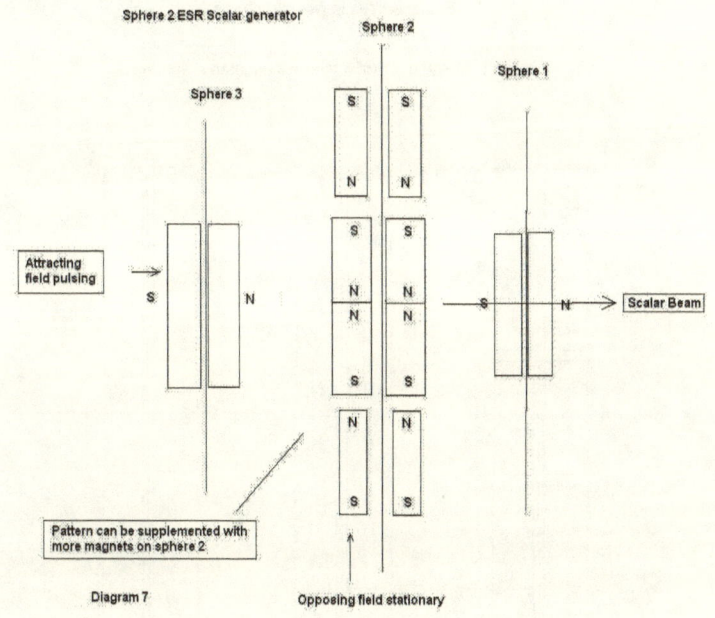

Diagram 7

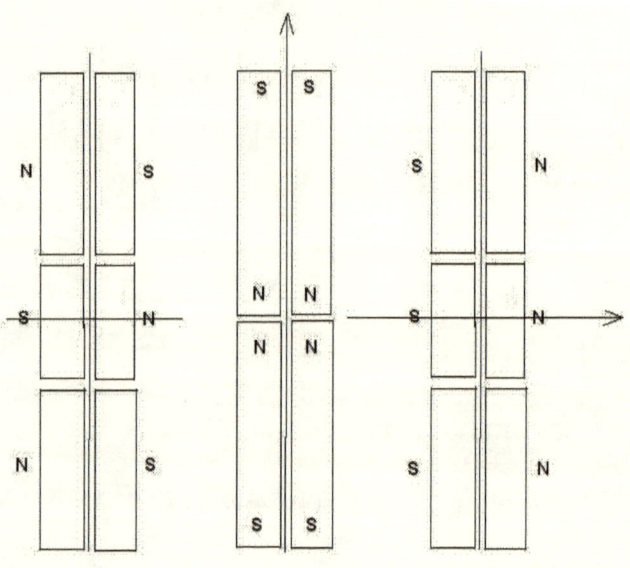

Diagram 8

Combining a balanced force with an unbalanced force

Unbalancing force is between spheres 1 an 3
Balanced force is between Spheres 2 - 3 and 2 - 1

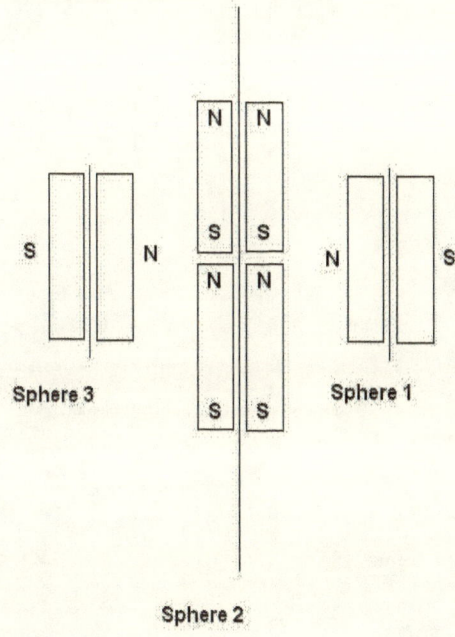

Diagram 9

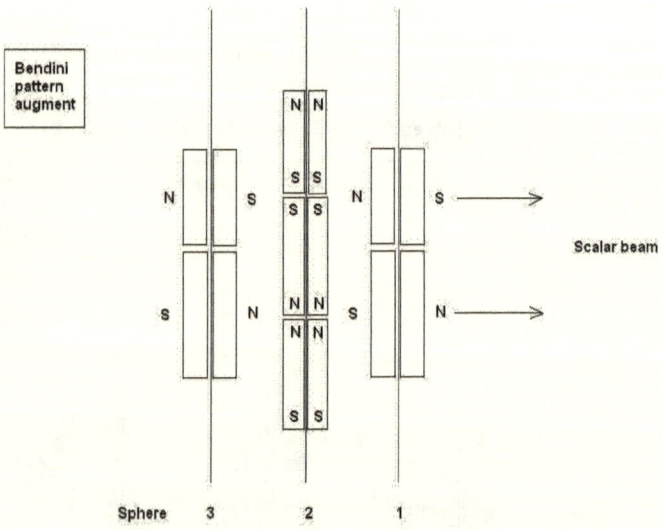

Diagram 10

Kosol and Koeun Noun Ouch Spherical Generator

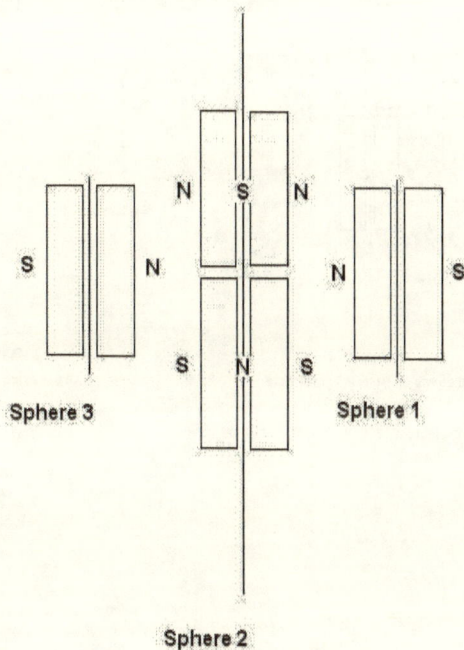

This all disc magnet layout may work as well to provide a host of scalar wave angles

Vertical field is a series of 3 virtual fields created between two magnets

Sphere 2 turns freely if all magnets are alligned perfectly

Diagram 11

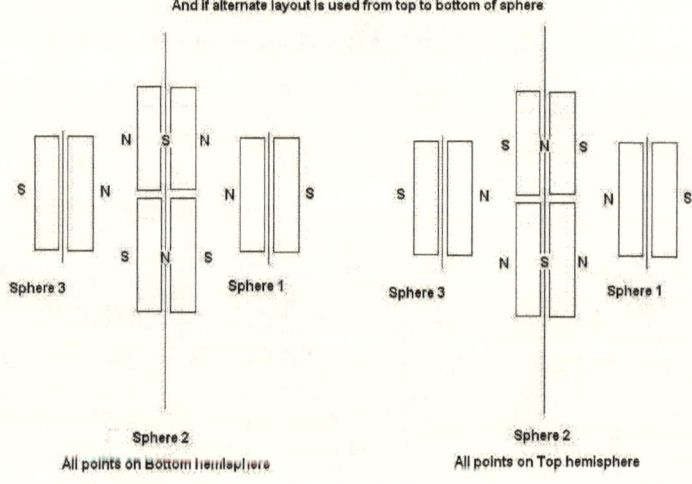

Diagram 12

Kosol and Koeun Noun Ouch Spherical Generator

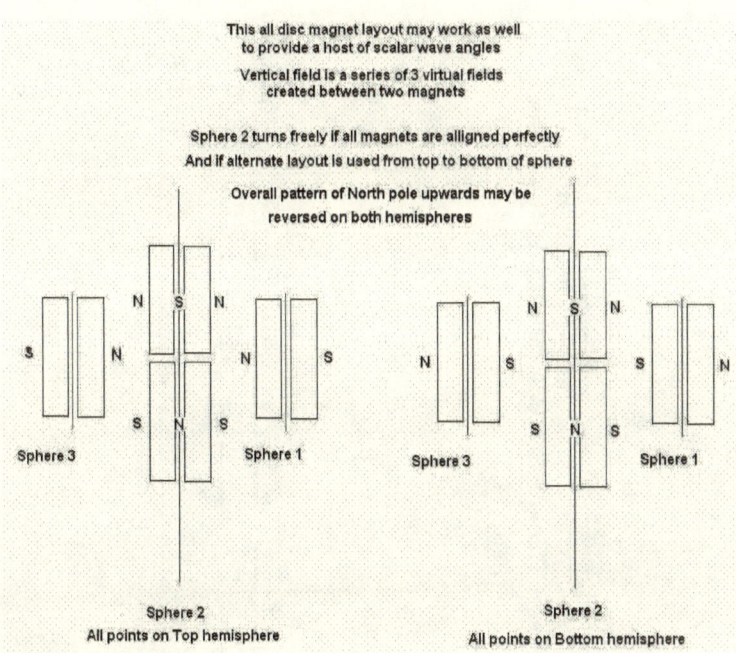

Diagram 13

sphere 5	Material	Diameter mm	Thickness mm	Thickness mm	Volume cm3	Material Density gr/cm3	Mass Kg
Layer 4	BatiO3 + Epoxi N-type	46.6	2.3	0	282	9.8	2.7636
Layer 3	NdFeB	46.6	4.5	0	271	2.7	0.7317
Layer 2	BatiO3 + Epoxi N-type	46.6	4.5	0	261	9.8	2.5578
Layer 1	NdFeB	46.6	4.5	0	231	9.8	2.2638
sphere set	all layers	248.57	218.36	0	1045		8.3169

	Material	Diameter	Thickness	Thickness	Volume	Material Density	Mass
		mm	mm	mm	cm3	gr/cm3	Kg
sphere 4	BatiO3 + Epoxi N-type	37.70105	1.878	0	282	9.8	2.7636
Layer 4	NdFeB	37.7	3.7083	0	271	2.7	0.7317
Layer 3	BatiO3 + Epoxi N-type						
Layer 2		37.7	3.7	0	261	9.8	2.5578
Layer 1	NdFeB	37.7	3.7	0	231	9.8	2.2638
sphere set	all layers	197.93	179.81	18.12	1045		8.3169

sphere 3	Material	Diameter mm	Thickness mm	Thickness mm	Volume cm3	Material Density gr/cm3	Mass Kg
Layer 4	BatiO3 + Epoxi N-type	30.55886	1.5	0	282	9.8	2.7636
Layer 3	NdFeB	30.5	3	0	271	2.7	0.7317
Layer 2	BatiO3 + Epoxi N-type	30.5	3	0	261	9.8	2.5578
Layer 1	NdFeB	30.5	3	0	231	9.8	2.2638
sphere set	all layers	161.27	144.32	16.95	1045		8.3169

sphere 2	Material	Diameter mm	Thickness mm	Thickness mm	Volume cm3	Material Density gr/cm3	Mass Kg
Layer 4	BatiO3 + Epoxi N-type	25	1.23	0	282	9.8	2.7636
Layer 3	NdFeB	25	2.5	0	271	2.7	0.7317
Layer 2	BatiO3 + Epoxi N-type	25	2.5	0	261	9.8	2.5578
Layer 1	NdFeB	25	2.5	0	231	9.8	2.2638
sphere set	all layers	131.61	115.61	16	1045		8.3169

sphere 1	Material	Diameter mm	Thickness mm	Thickness mm	Volume cm3	Material Density gr/cm3	Mass Kg
Layer 4	BatiO3 + Epoxi N-type	20	1	0	282	9.8	2.7636
Layer 3	NdFeB	20	2	0	271	2.7	0.7317
Layer 2	BatiO3 + Epoxi N-type	20	2	0	261	9.8	2.5578
Layer 1	NdFeB	20	2	0	231	9.8	2.2638
Magnet cup	all layers	108	92	16	1045		8.3169

Kosol and Koeun Noun Ouch Spherical Generator

CORE	Material	External radius mm	Internal radius mm	Thickness mm	Volume aprox cm3	Material Density gr/cm3	Mass Kg
Layer 5	Barium Titanate	62.57	60.57	2	92	BaTiO3 75% 5,8 gr/cm3 Epoxi 25% 1,7 gr/cm3	0.4393
Layer 4	Bismuth	60.57	58.57	2	86	9.8	0.8428
Layer 3	Alluminium	58.57	56.57	2	80	2.7	0.216
Layer 2	Bismuth	56.57	54.57	2	75	9.8	0.735
Layer 1	Barium Titanate Epoxi N-type	54.57	50.57	4	133	BaTiO3 75% 5,8 gr/cm3 Epoxi 25% 1,7 gr/cm3	0.635075
CORE rings magnet 50%	Magnet	50.57	0	50.57	270	7.5	2.025
CORE rings dielec 50%	Barium Titanate Epoxi N-type	50.57	0	50.57	270	BaTiO3 75% 5,8 gr/cm3 Epoxi 25% 1,7 gr/cm3	1.28925
sphere set	all layers	62.57	0	62.57	1006		6.182425

Total bismuth	47.766925		
mass shaft			
sph 1 3 5	24.9507		
sph C 2 4	22.816225		46*79/100
		36.34	60.30534

Rectangular inductance is computed as a degenerated conical inductace whose a side of is of D length and R is D*sqrt (2) length respectively.
W is the 'depth' of the coil
the formula is :
(((N * R) * (N * R)) / ((8 * R) + (11 * W)))-D

Remember that when you feed a coil with current always a counter tension is raised according the the formula :

$$VL = -L_{eq} \frac{d\, i(t)}{dt}$$

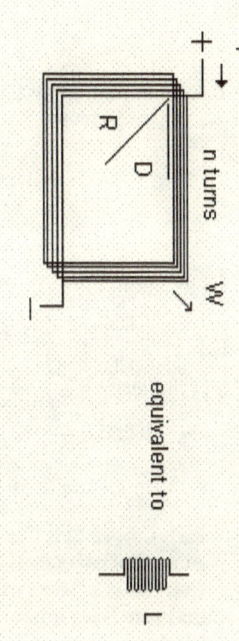

equivalent to

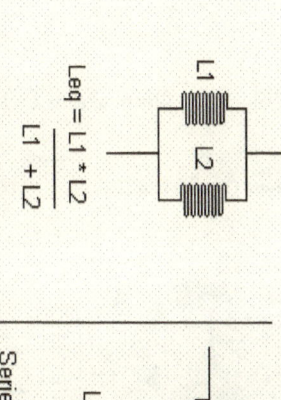

Serie of 2 inductance ,
the general formula is
the equivalent inductance is
the sum of inductances

Leq = L1+ L2

L1   L2

Parallel of 2 inductance
the general formula is
the equivalent inductance is
the reciprocal of the sum of reciprocals

$$Leq = \frac{L1 * L2}{L1 + L2}$$

L1
L2

Kosol Multi Layer Cone Coil Device. [2005]   **Theoretical Drawing ONLY.**

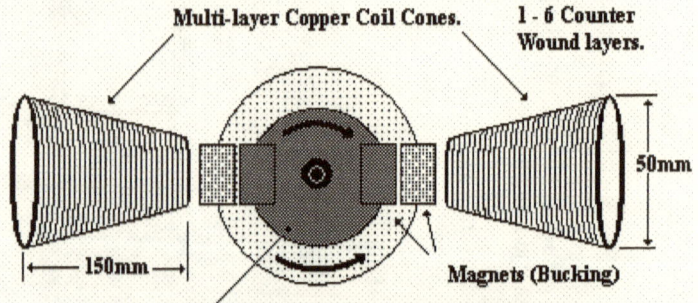

Multi-layer Copper Coil Cones.   1 - 6 Counter Wound layers.

150mm   50mm

Magnets (Bucking)

Centre Shaft varies speed for different effects.

Each Copper Coil Cone windings are wound in alternate layers going Clockwise then Counter Clockwise. Each Pair of Copper Coil Cones must be wire wound opposite to the other. Ie. The very first wire layer is wound Clockwise on one cone and Counter Clockwise on the other.

Theoretical Drawing ONLY.   Drawing G.D.Mutch

Kosol and Koeun Noun Ouch Spherical Generator

Kosol Ouch, Koeun Noun Ouch and David Lowrance

Kosol and Koeun Noun Ouch Spherical Generator

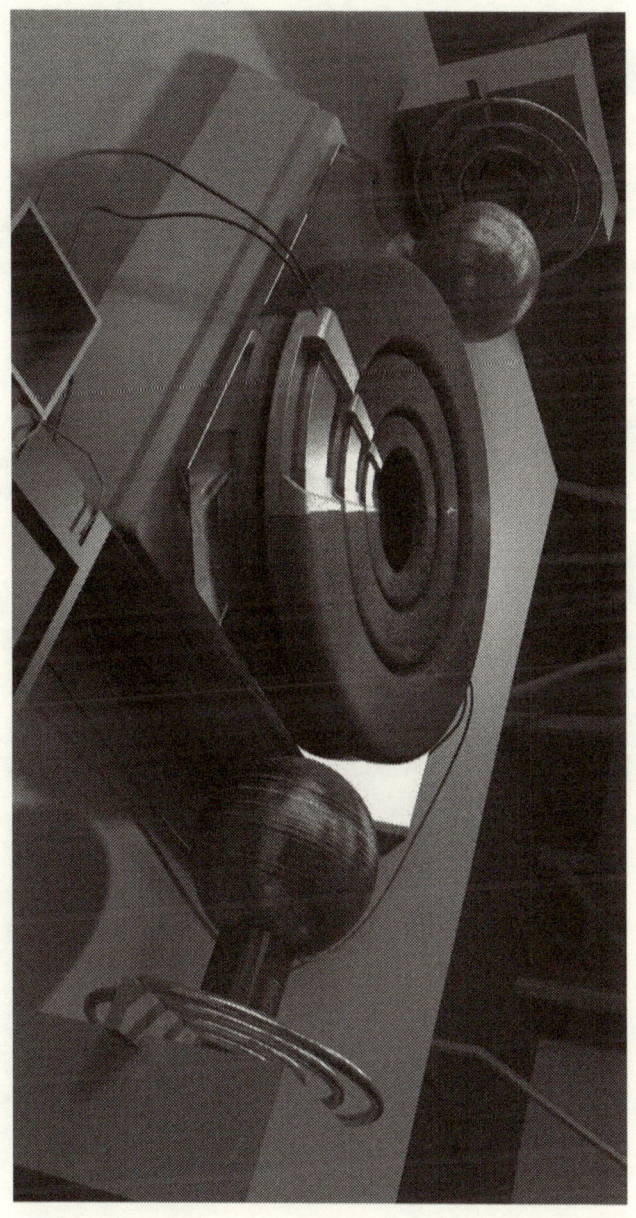

Kosol Ouch, Koeun Noun Ouch and David Lowrance

Kosol and Koeun Noun Ouch Spherical Generator

Kosol Ouch, Koeun Noun Ouch and David Lowrance

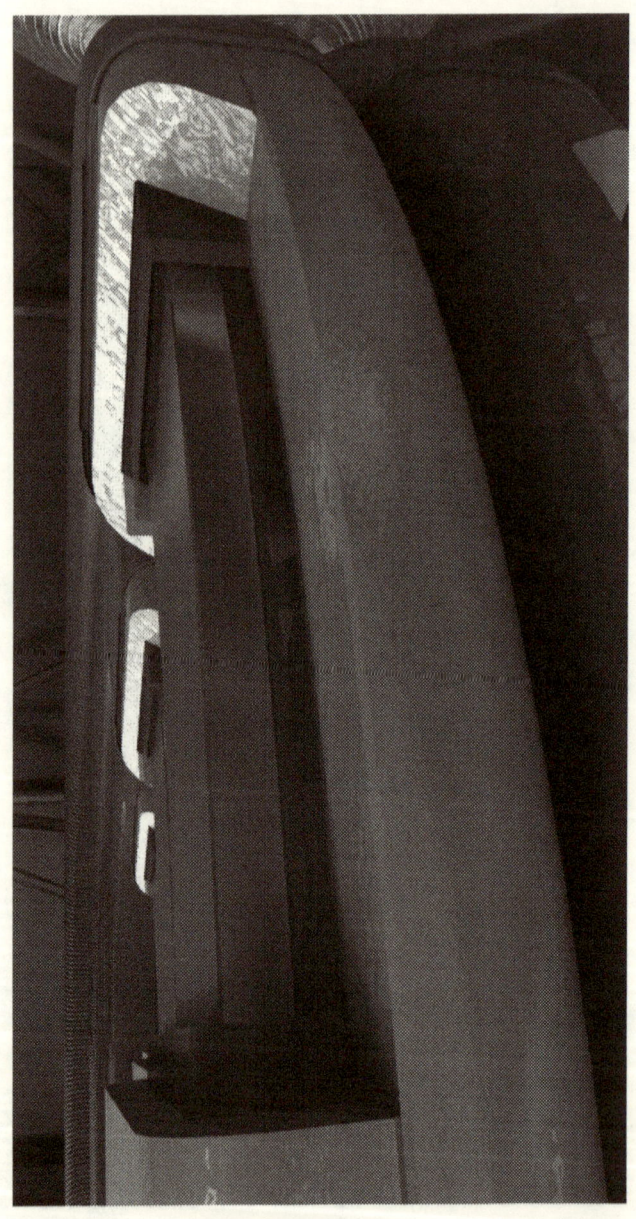

Kosol and Koeun Noun Ouch Spherical Generator

Kosol Ouch, Koeun Noun Ouch and David Lowrance

Kosol and Koeun Noun Ouch Spherical Generator

Kosol Ouch, Koeun Noun Ouch and David Lowrance

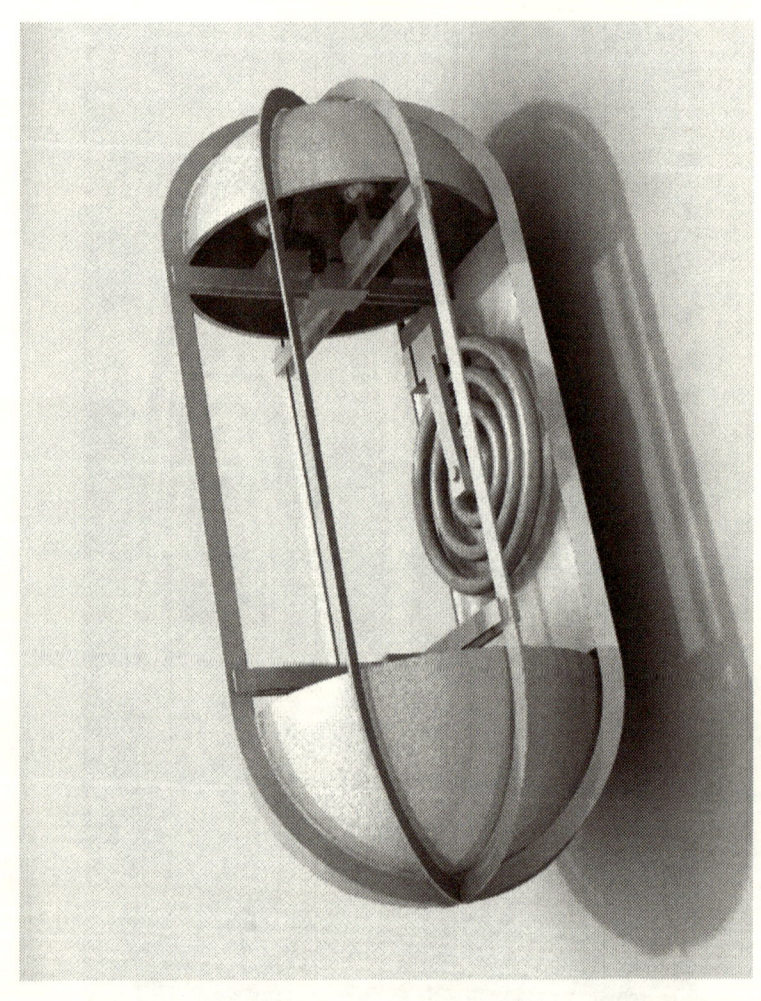

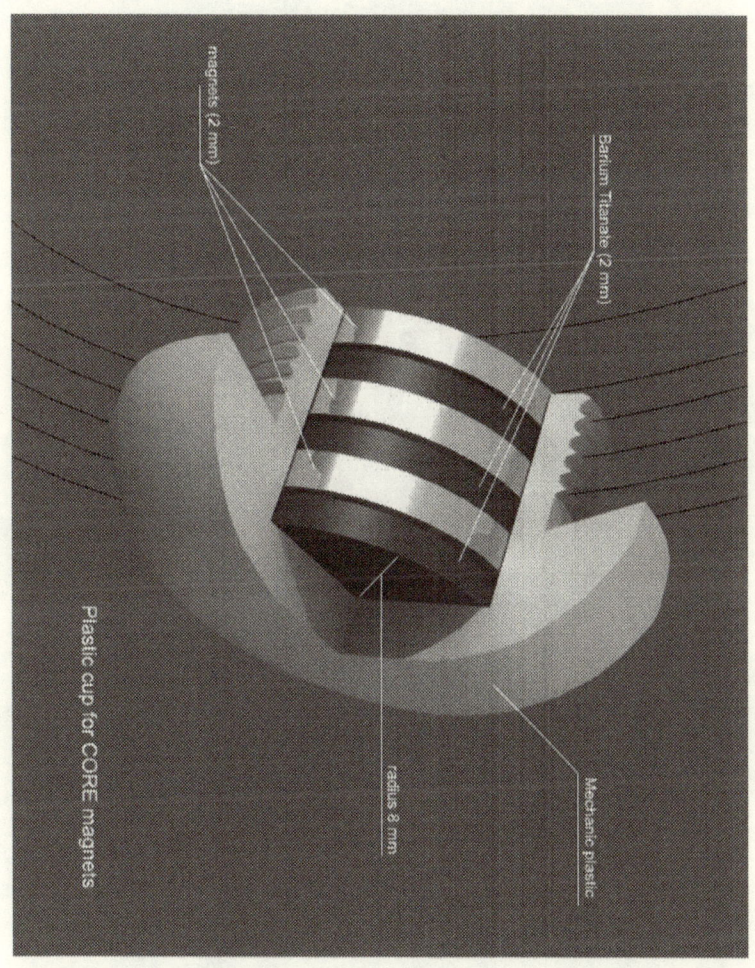

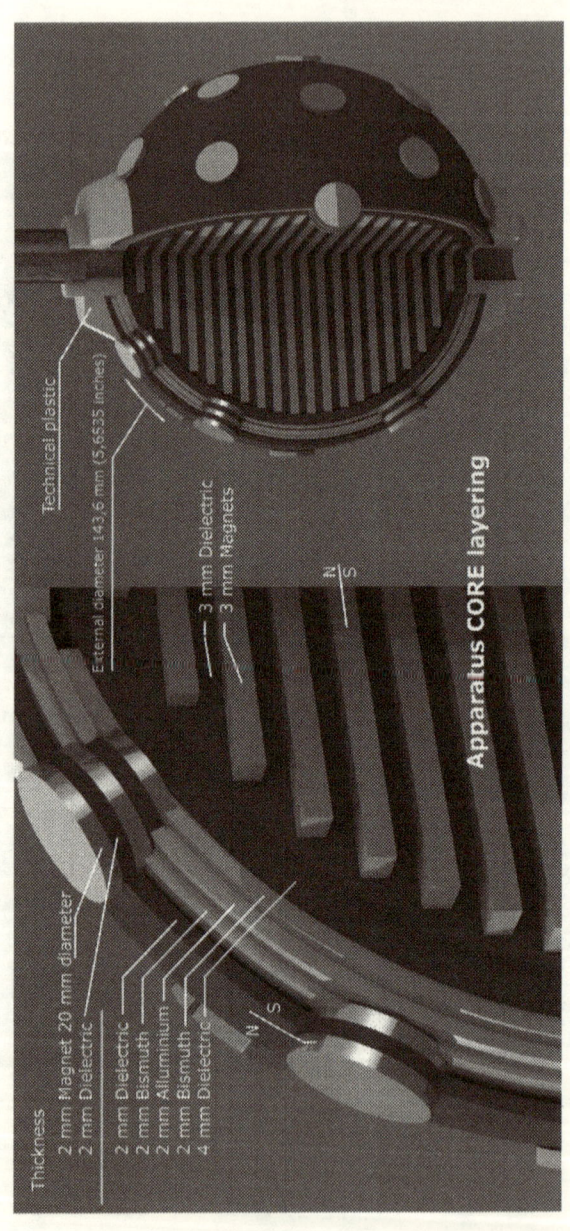

Kosol and Koeun Noun Ouch Spherical Generator

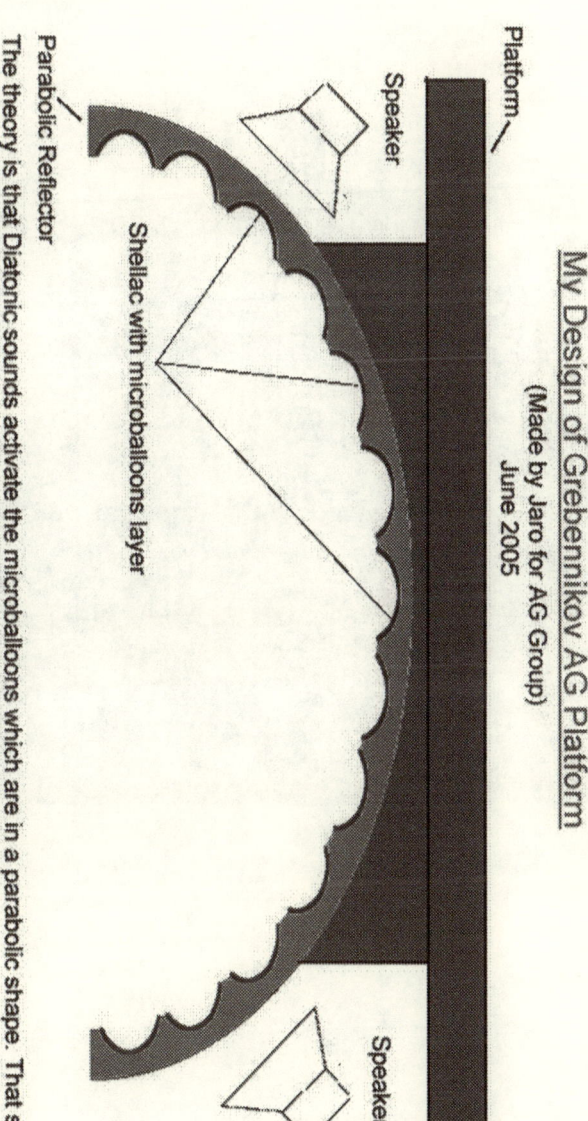

My Design of Grebennikov AG Platform
(Made by Jaro for AG Group)
June 2005

Platform

Speaker

Shellac with microballoons layer

Parabolic Reflector

The theory is that Diatonic sounds activate the microballoons which are in a parabolic shape. That should produce torsion fields strong enough to block the downward flow of Aether.

Speaker

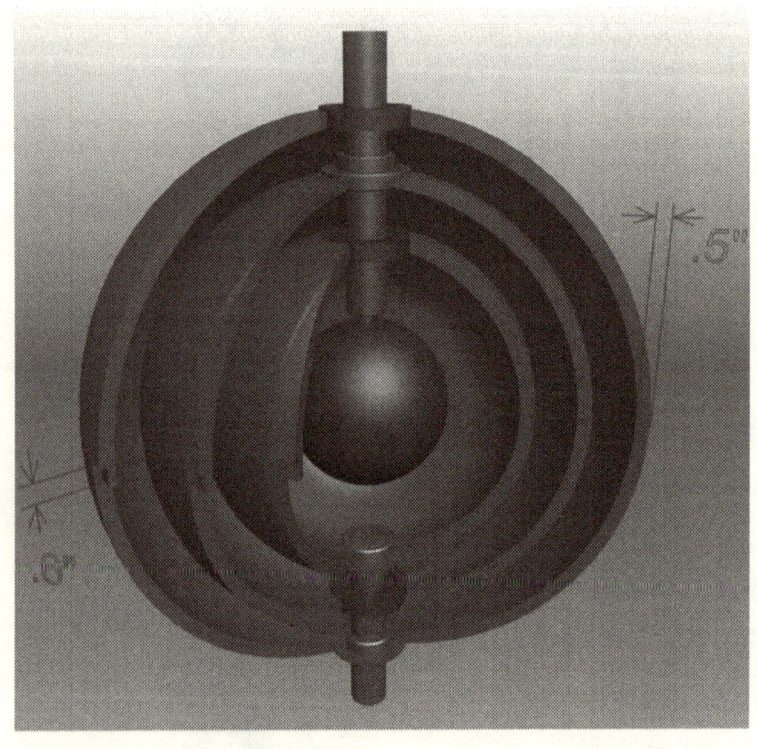

Kosol and Koeun Noun Ouch Spherical Generator

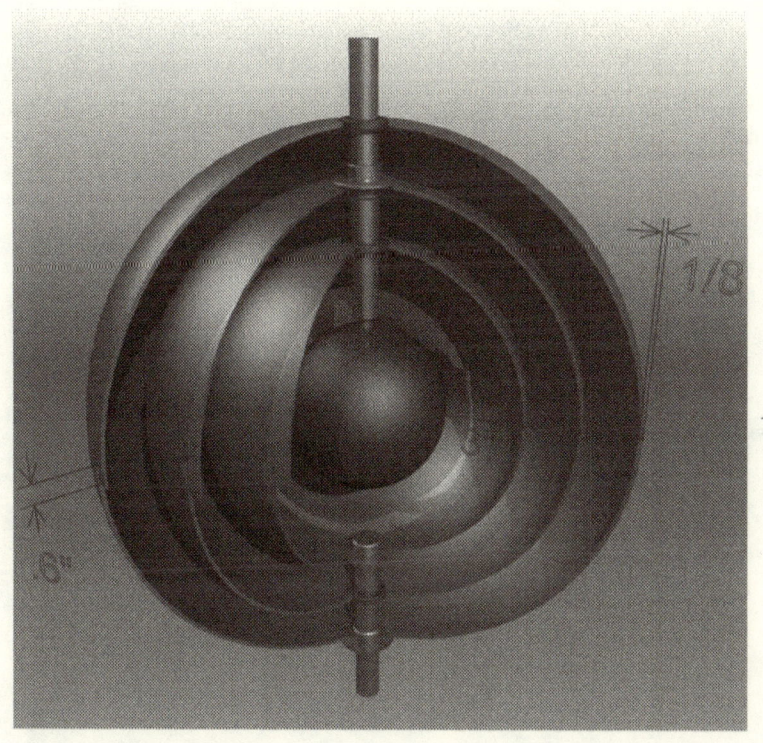

Kosol Ouch, Koeun Noun Ouch and David Lowrance

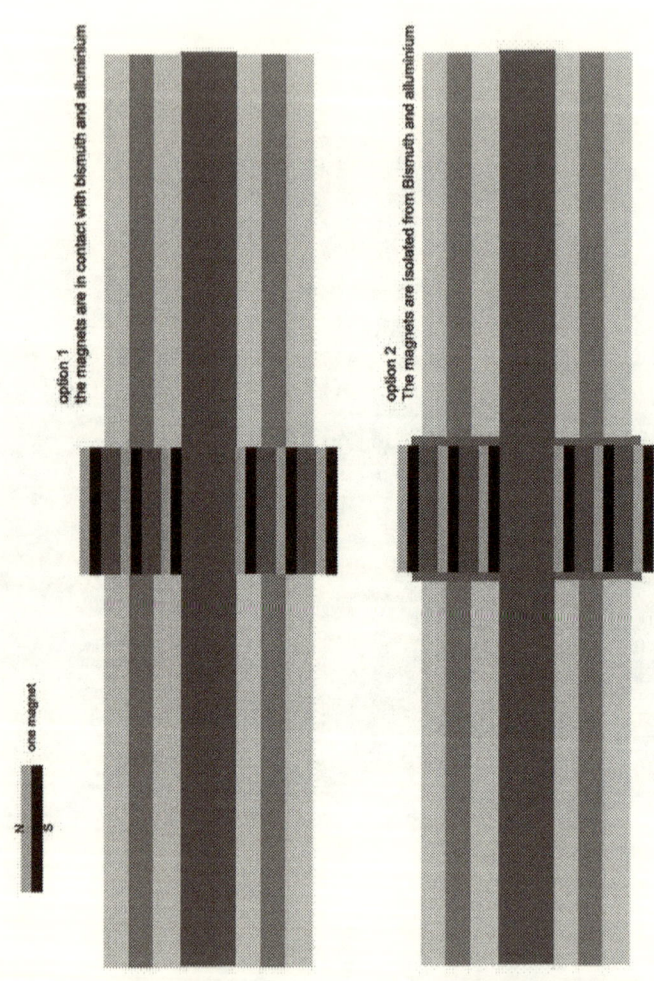

Kosol and Koeun Noun Ouch Spherical Generator

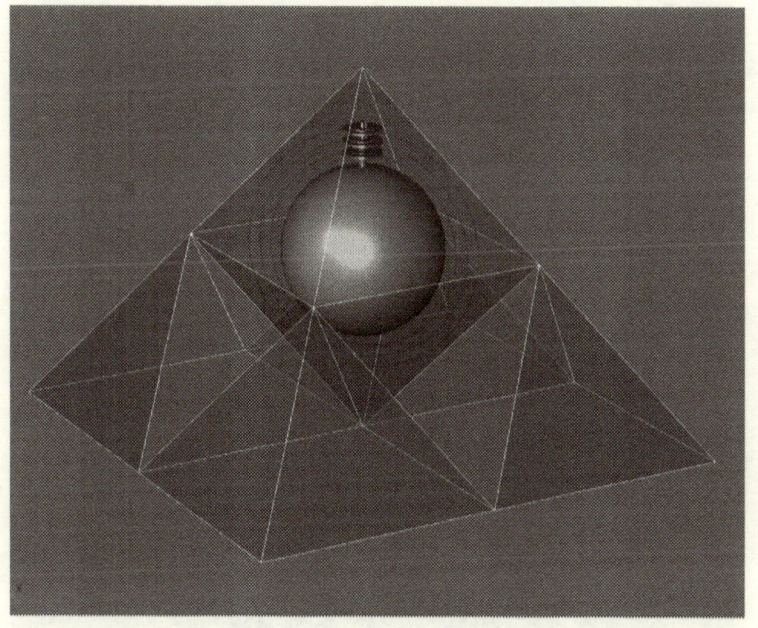

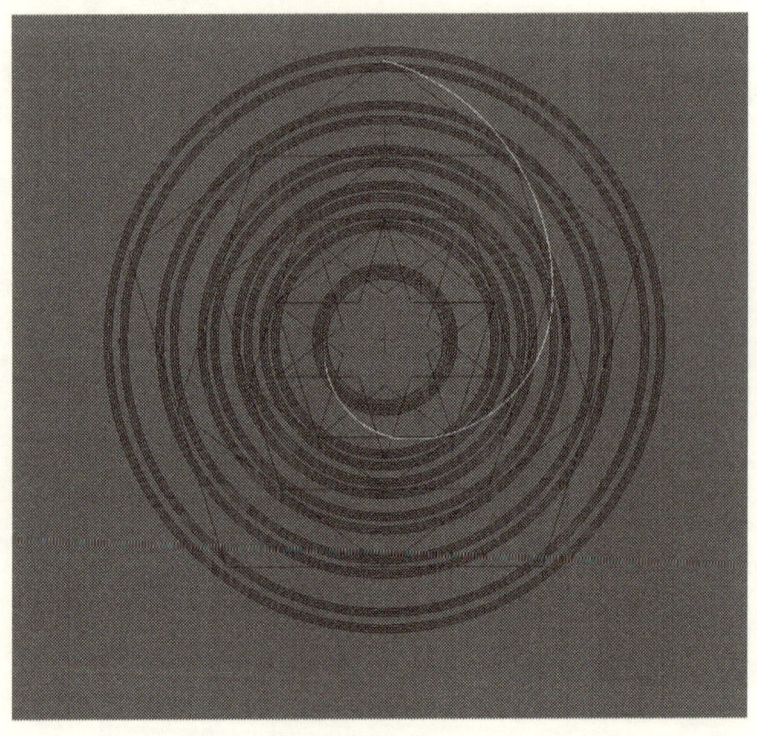

Kosol and Koeun Noun Ouch Spherical Generator

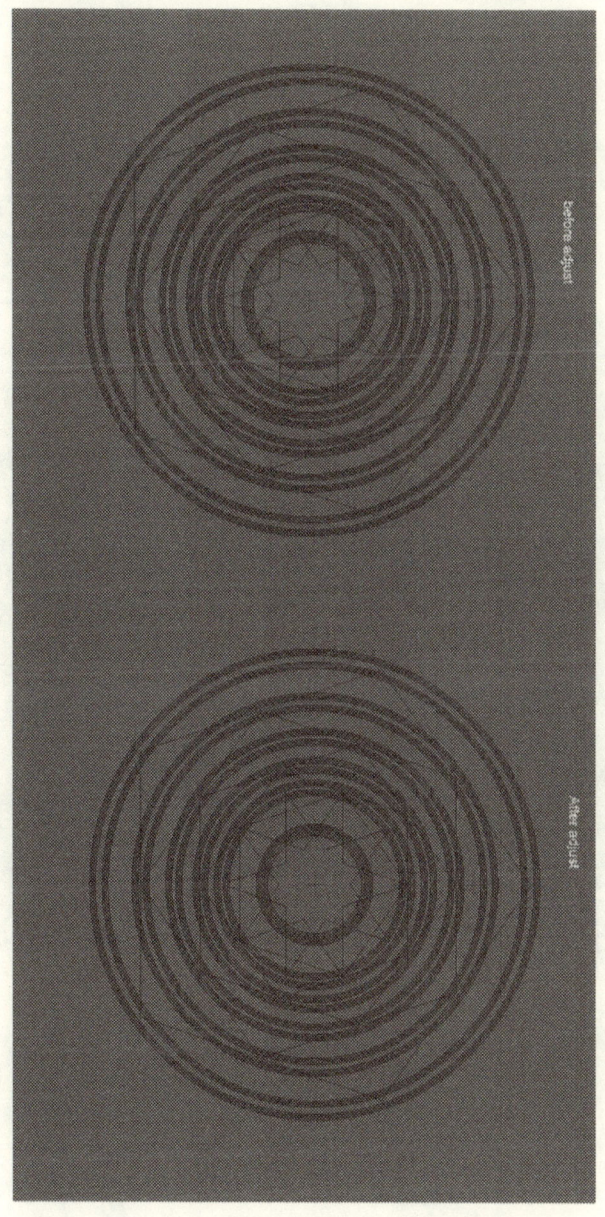

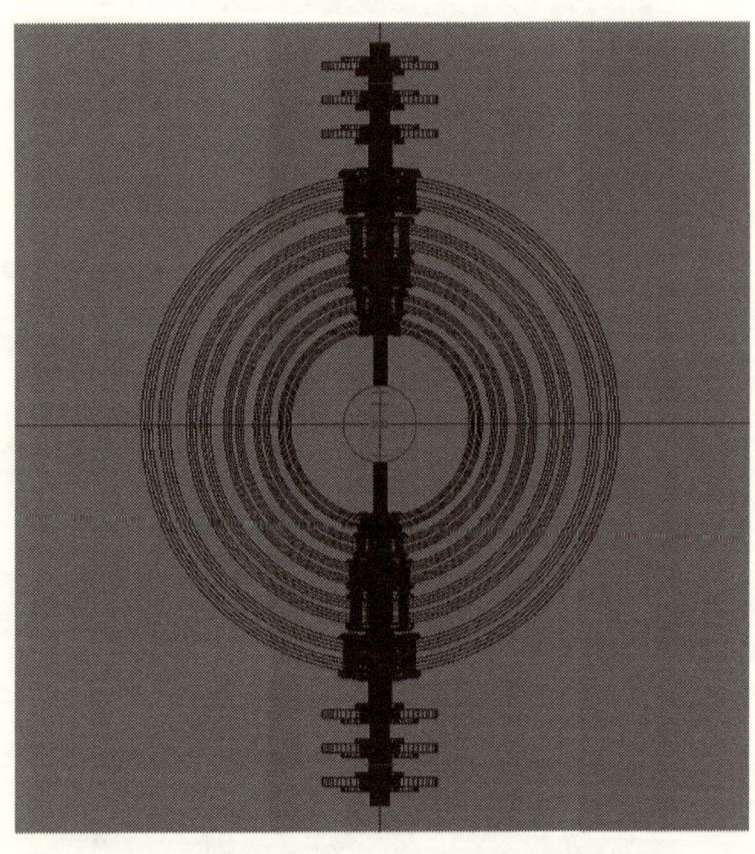

Kosol Ouch, Koeun Noun Ouch and David Lowrance

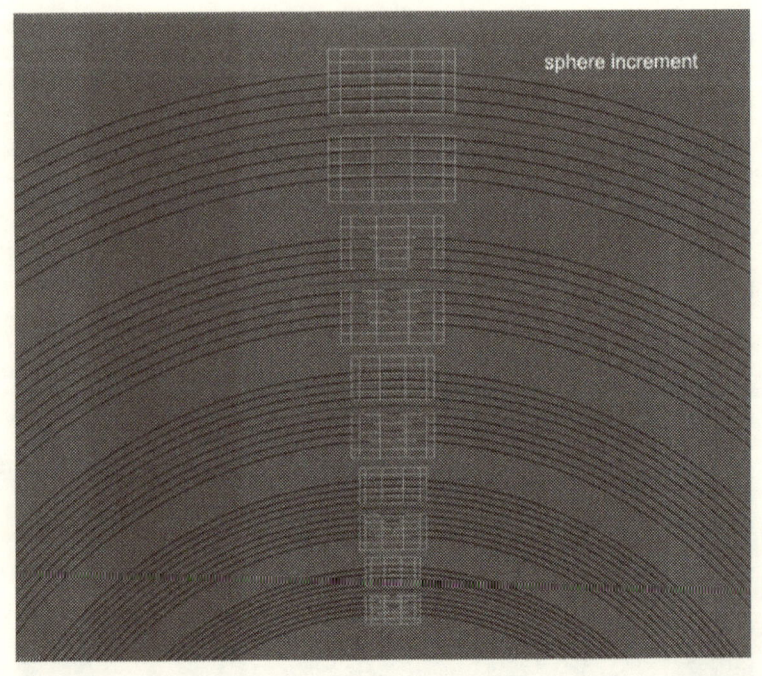

Kosol and Koeun Noun Ouch Spherical Generator

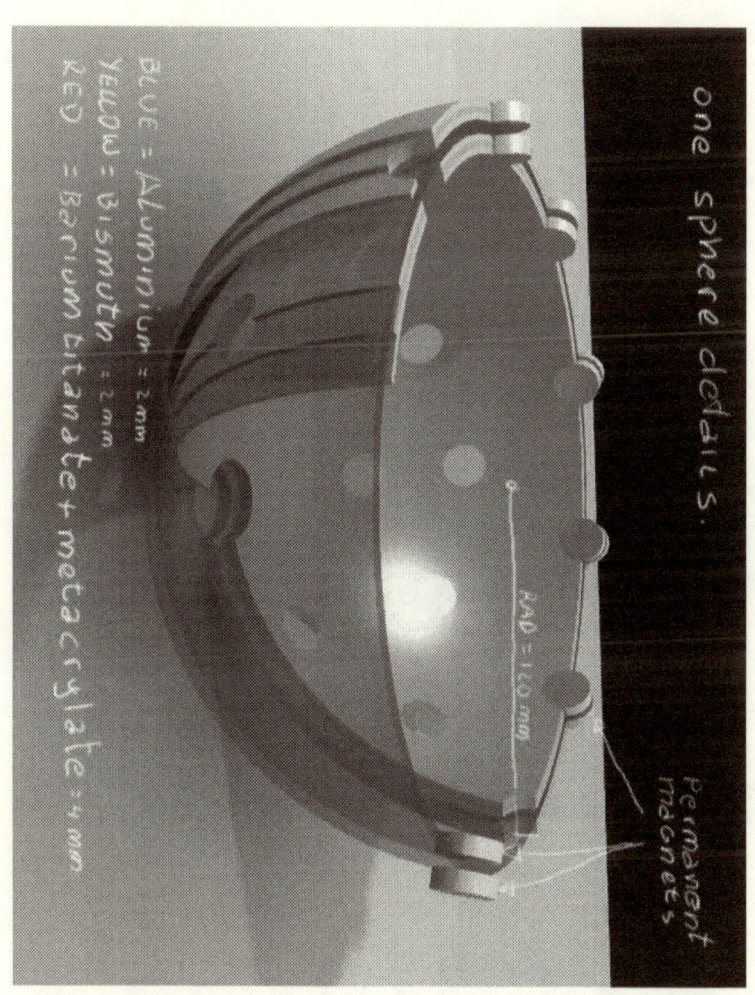

Kosol and Koeun Noun Ouch Spherical Generator

Kosol and Koeun Noun Ouch Spherical Generator

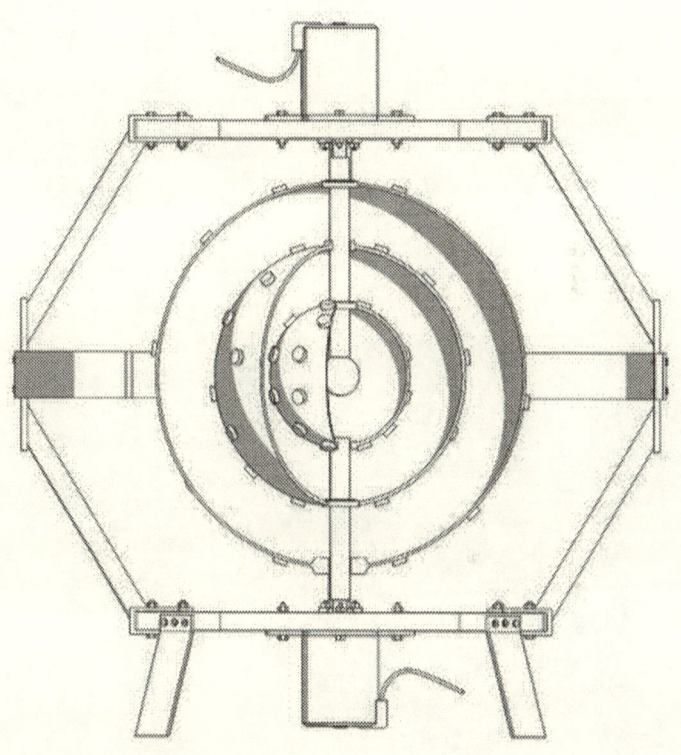

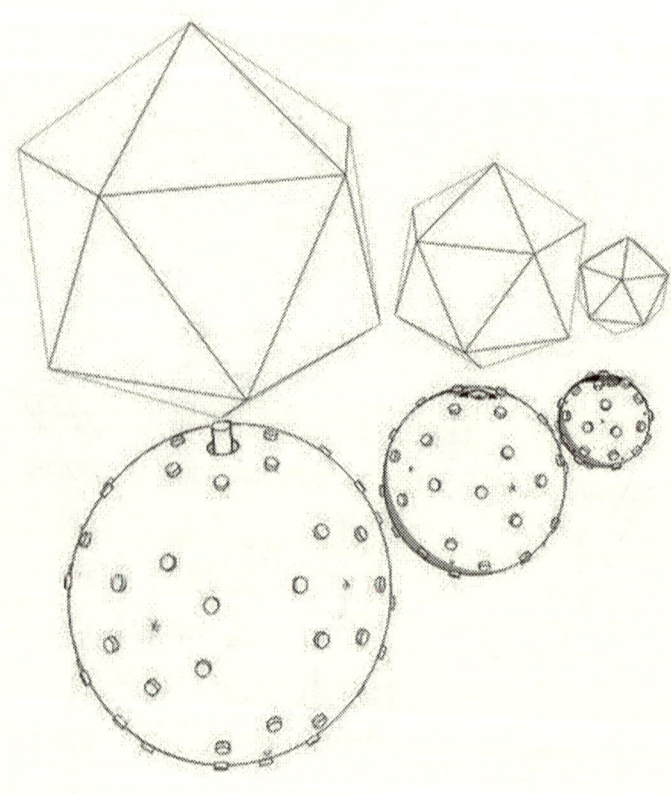

Kosol and Koeun Noun Ouch Spherical Generator

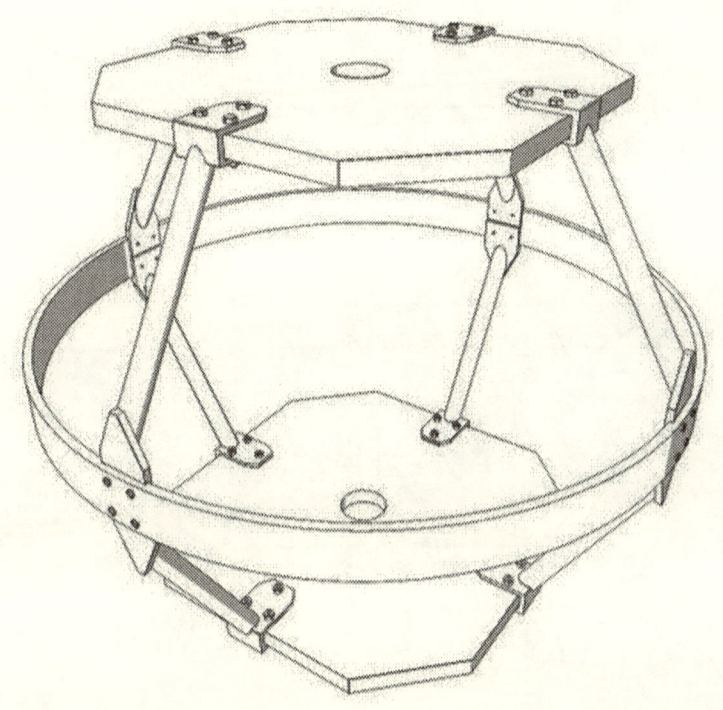

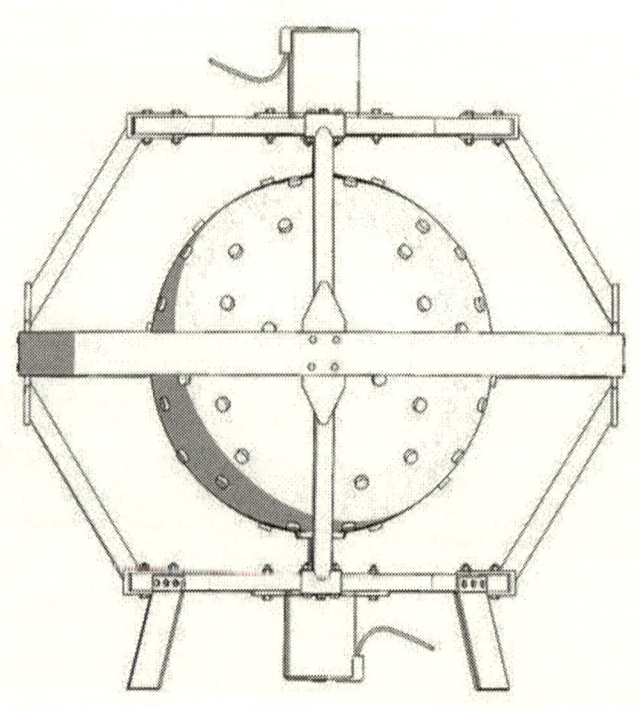

Kosol and Koeun Noun Ouch Spherical Generator

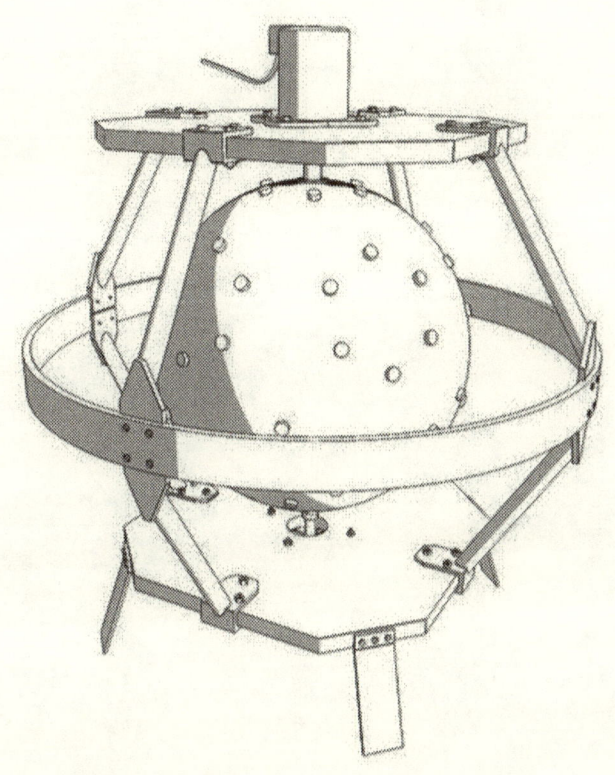

Kosol Ouch, Koeun Noun Ouch and David Lowrance

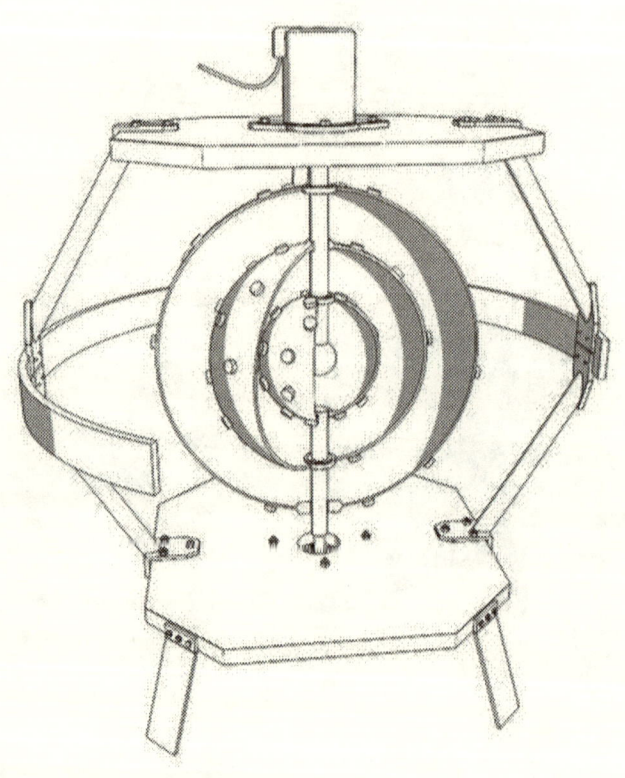

Kosol and Koeun Noun Ouch Spherical Generator

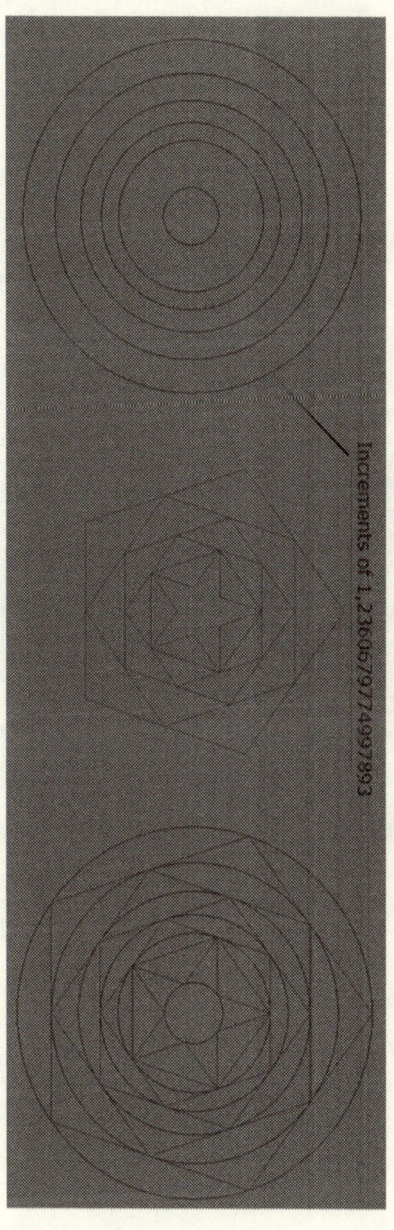

Increments of 1.2360679774997993

Kosol Ouch, Koeun Noun Ouch and David Lowrance

Kosol and Koeun Noun Ouch Spherical Generator

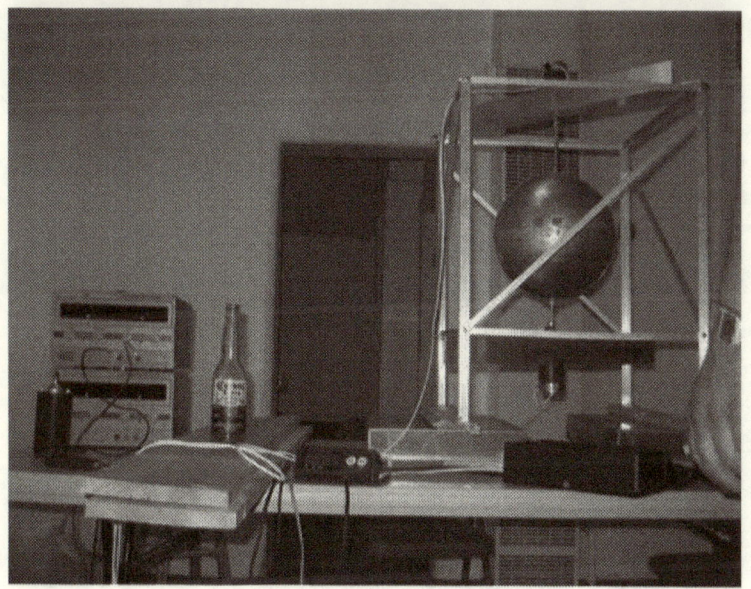

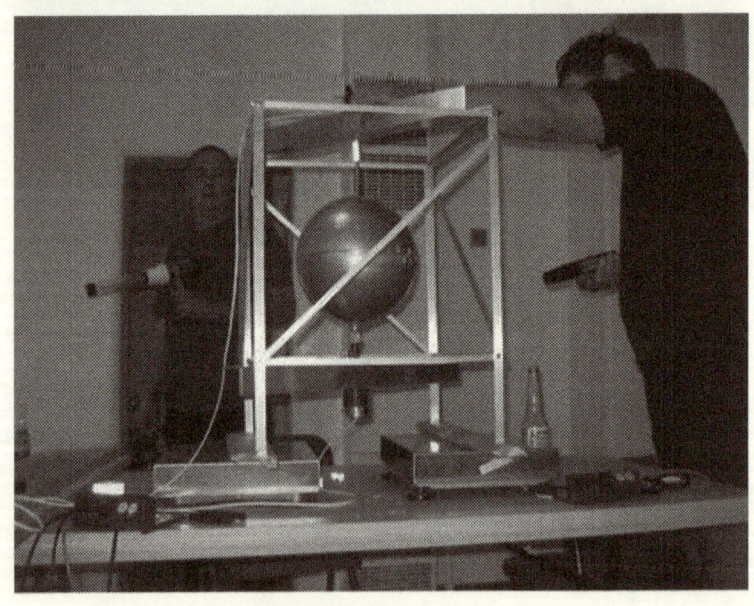

Kosol and Koeun Noun Ouch Spherical Generator

Kosol Ouch, Koeun Noun Ouch and David Lowrance

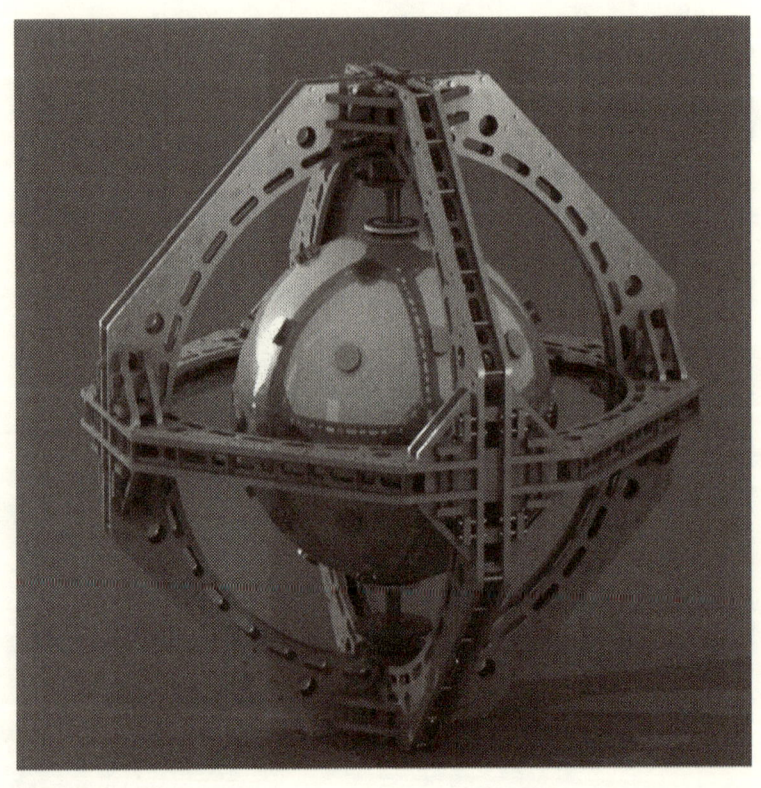

Kosol and Koeun Noun Ouch Spherical Generator

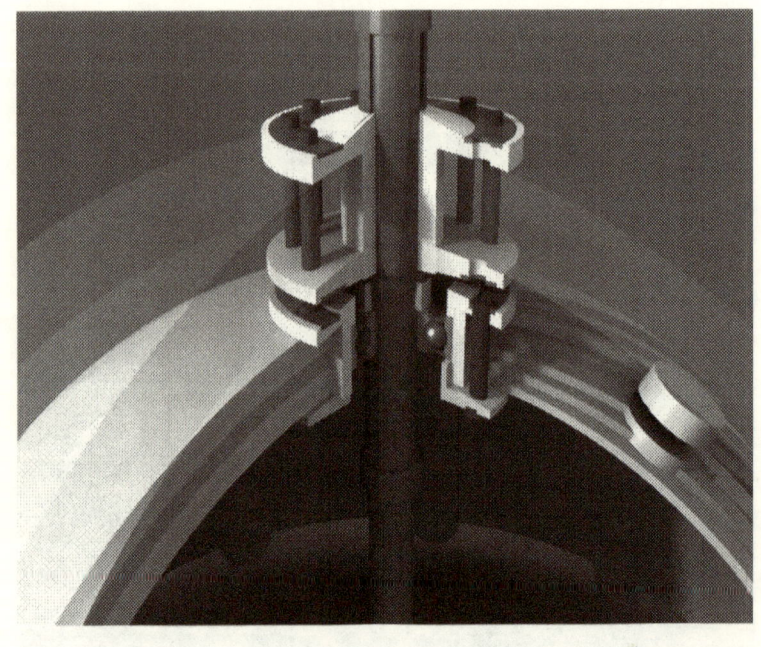

Kosol and Koeun Noun Ouch Spherical Generator

Kosol and Koeun Noun Ouch Spherical Generator

Kosol and Koeun Noun Ouch Spherical Generator

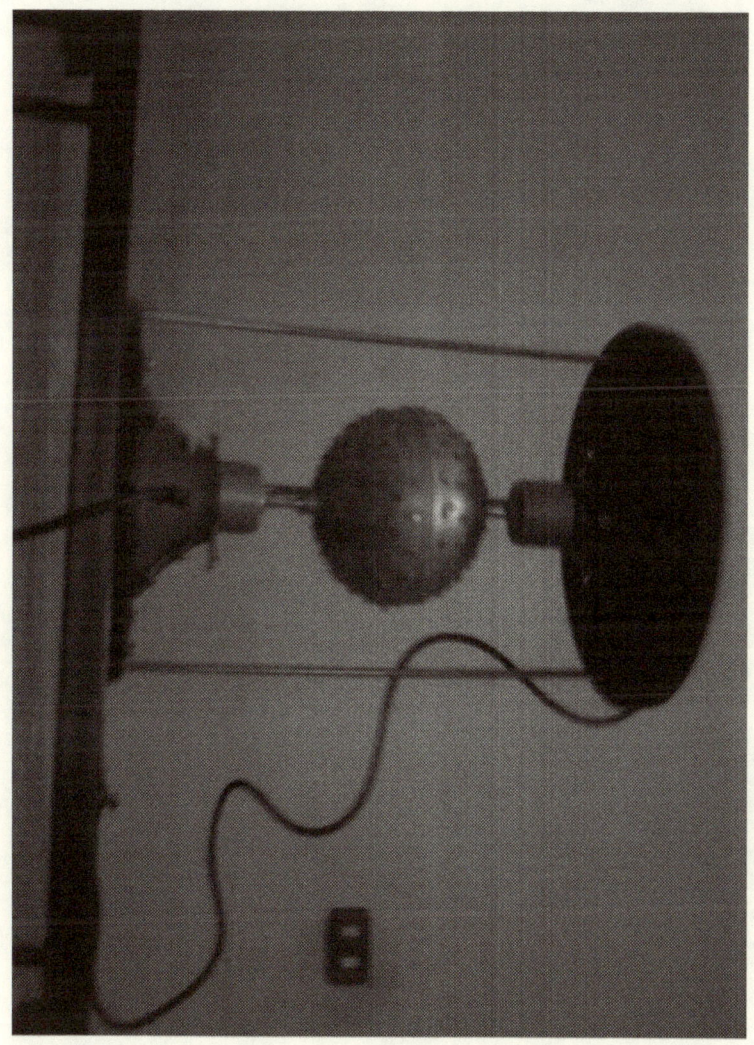

Kosol and Koeun Noun Ouch Spherical Generator

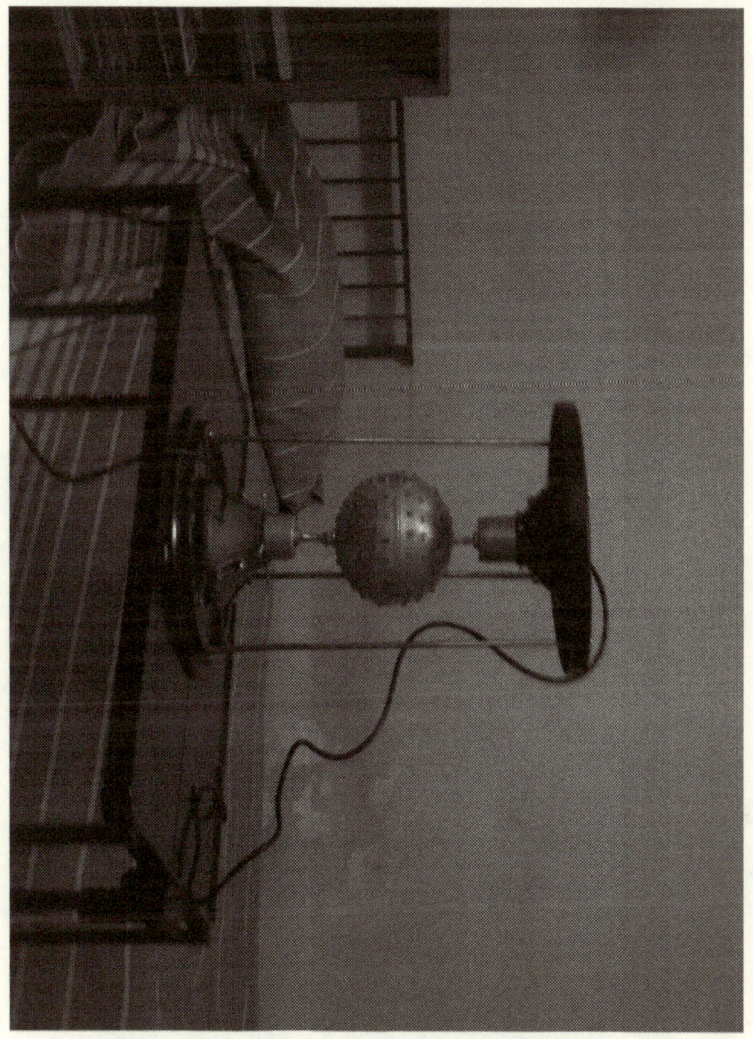

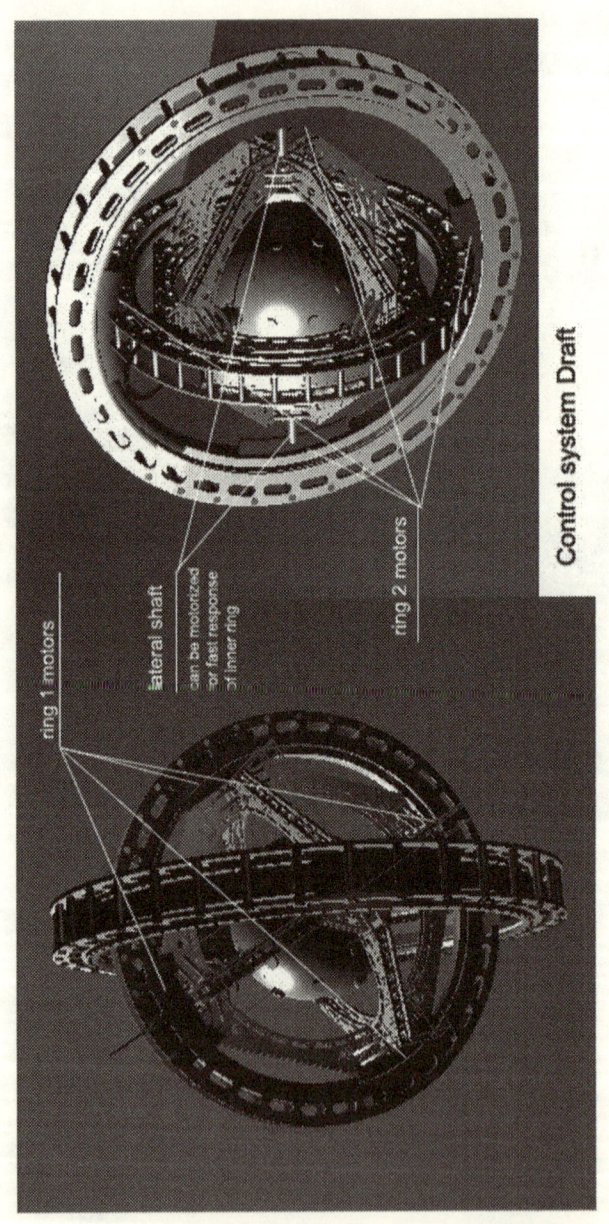

Kosol and Koeun Noun Ouch Spherical Generator

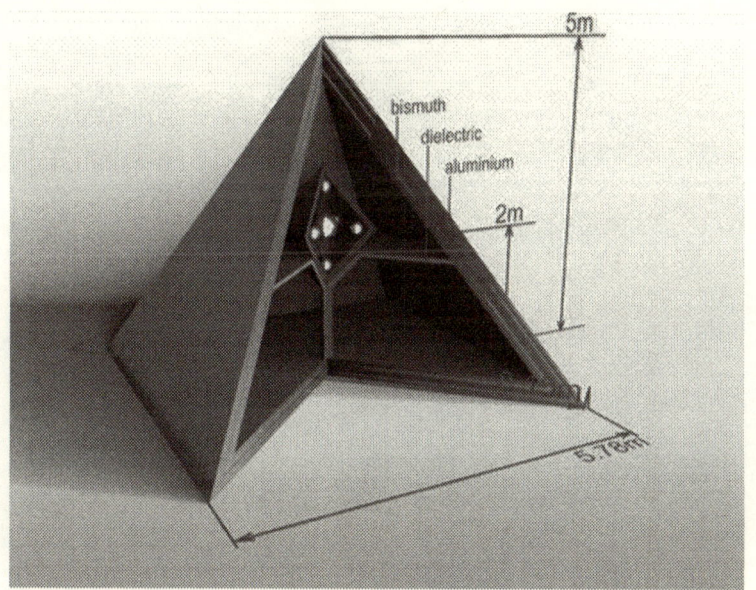

Kosol Ouch, Koeun Noun Ouch and David Lowrance

Kosol and Koeun Noun Ouch Spherical Generator

Kosol Ouch, Koeun Noun Ouch and David Lowrance

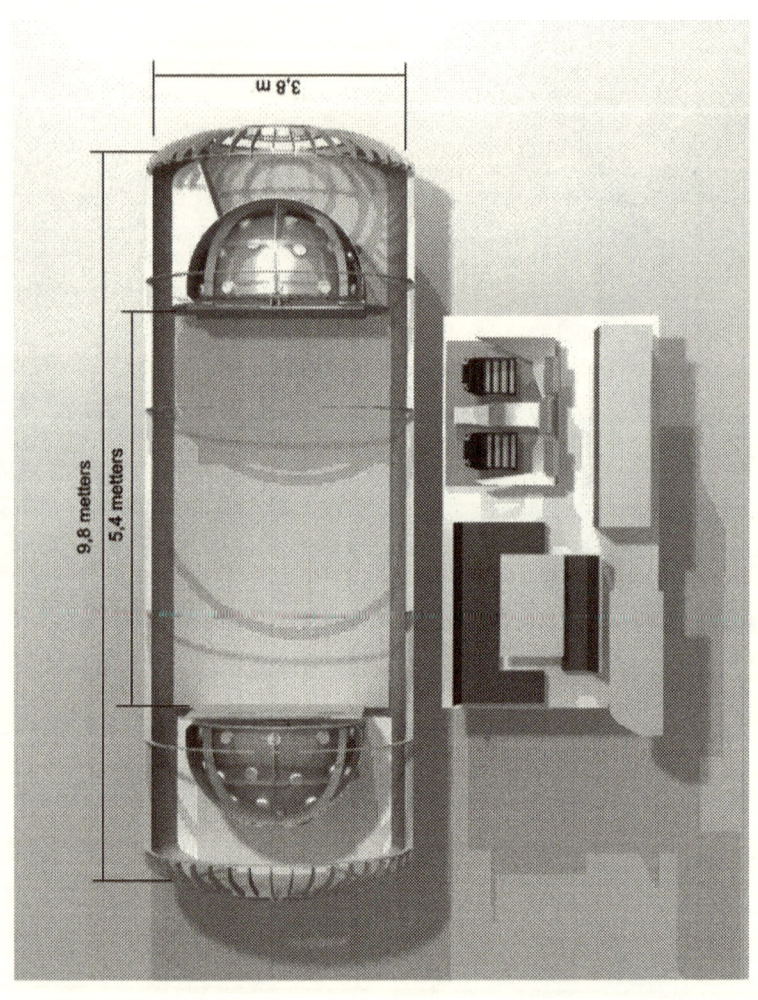

Kosol and Koeun Noun Ouch Spherical Generator

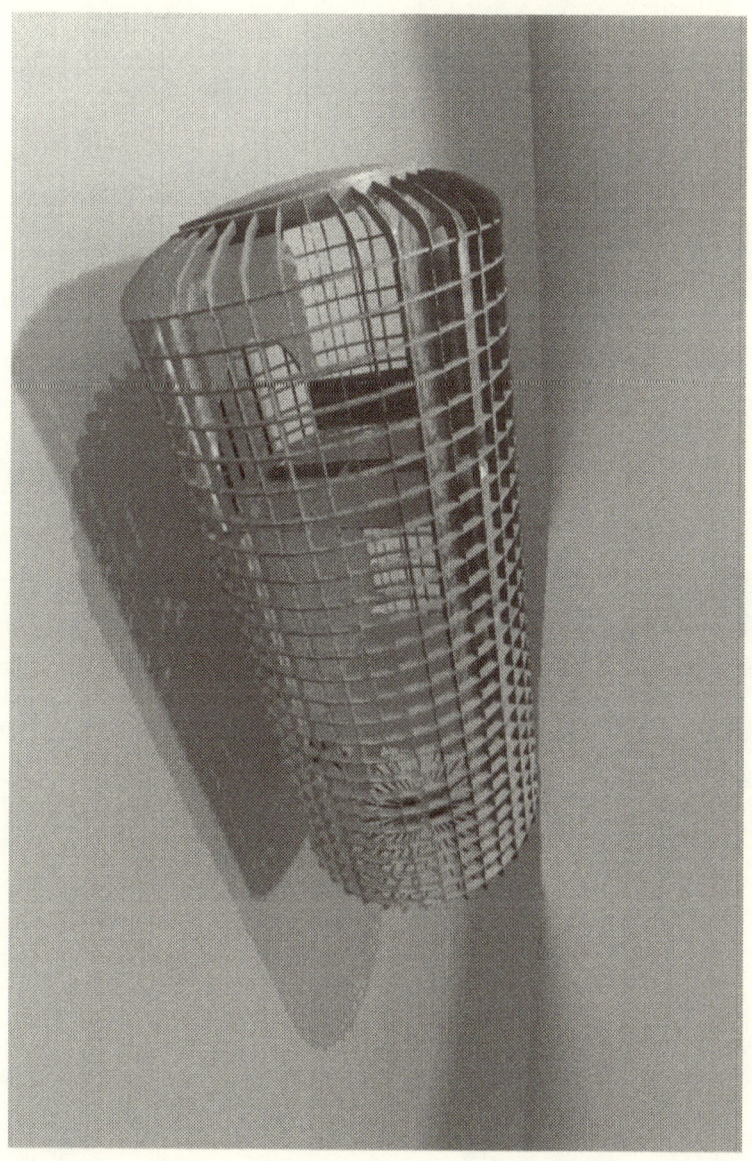

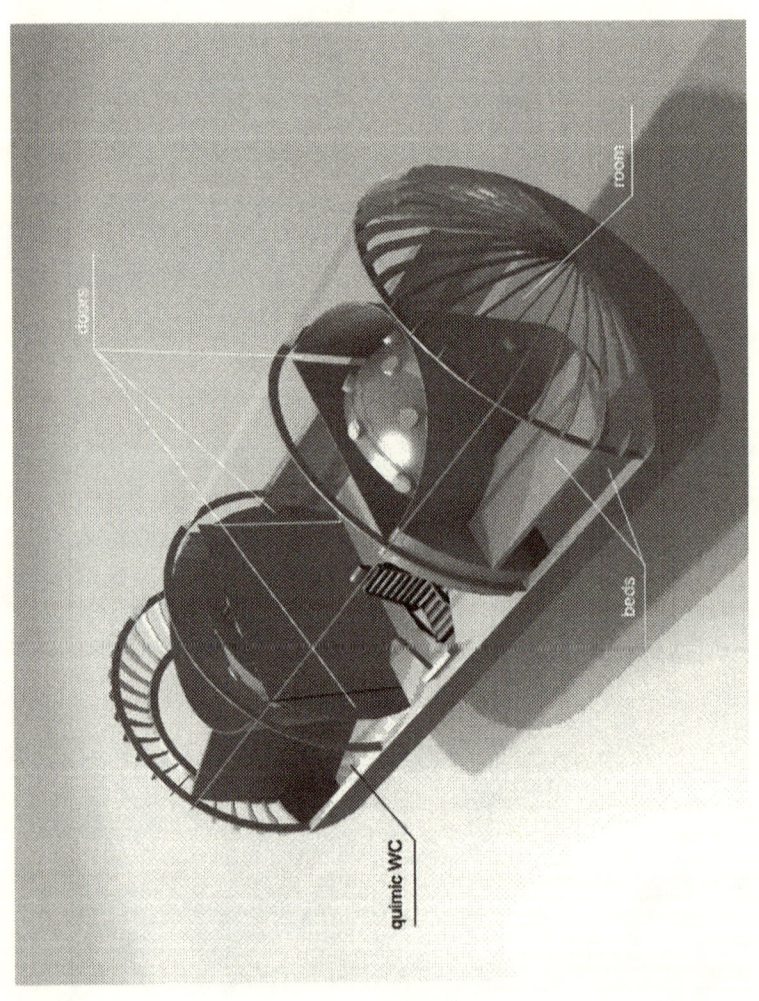

Kosol and Koeun Noun Ouch Spherical Generator

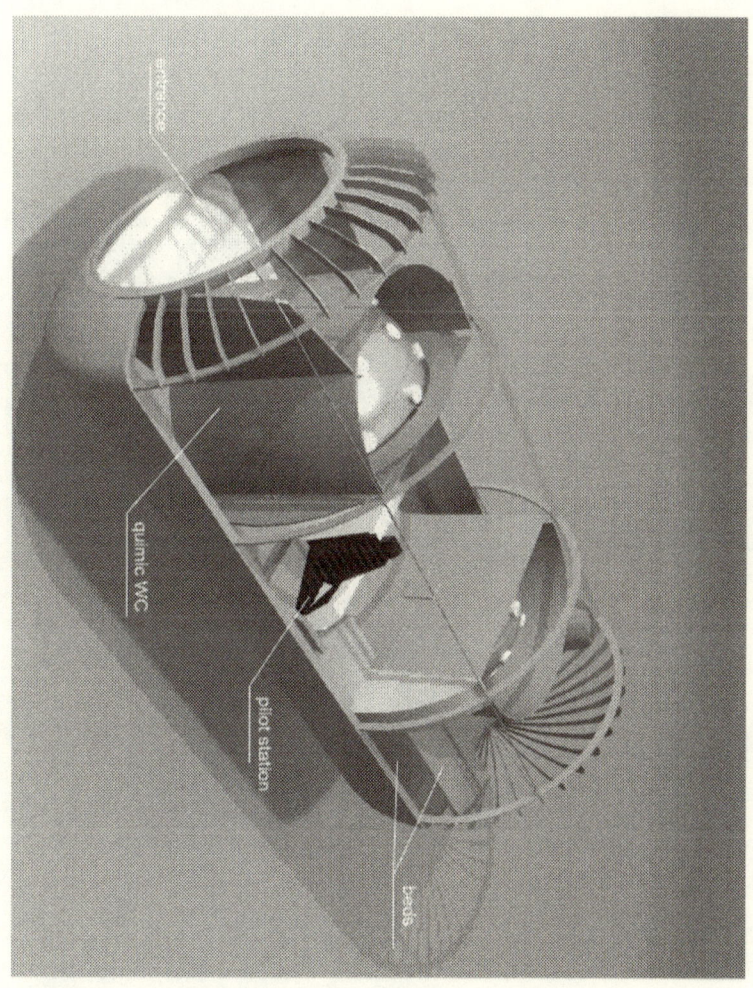

Kosol Ouch, Koeun Noun Ouch and David Lowrance

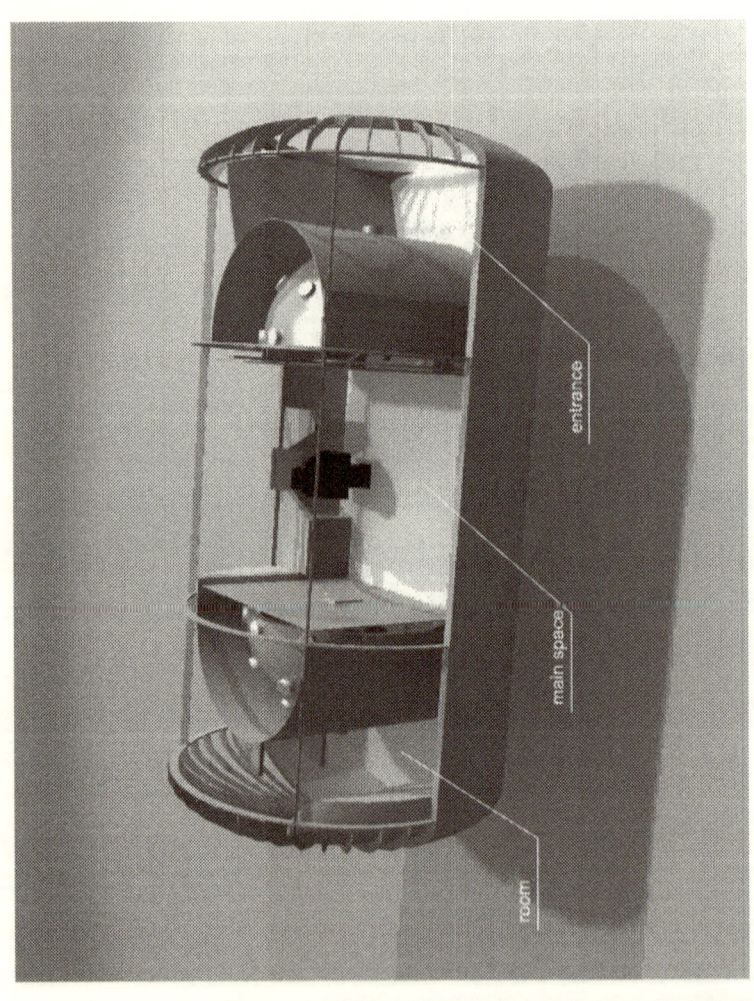

Kosol and Koeun Noun Ouch Spherical Generator

Kosol and Koeun Noun Ouch Spherical Generator

Kosol Ouch, Koeun Noun Ouch and David Lowrance

Kosol and Koeun Noun Ouch Spherical Generator

Kosol Ouch, Koeun Noun Ouch and David Lowrance

www.ingramcontent.com/pod-product-compliance
Lightning Source LLC
Chambersburg PA
CBHW030935180526
45163CB00002B/575